LIFE IN THE FASTING LANE

LIFE IN THE
FASTING
LANE

How to Make Intermittent Fasting a Lifestyle—and
Reap the Benefits of Weight Loss and Better Health

Jason Fung, MD | Eve Mayer | Megan Ramos

HARPER WAVE

An Imprint of HarperCollinsPublishers

This book contains advice and information relating to health care. It should be used to supplement rather than replace the advice of your doctor or another trained health professional. If you know or suspect you have a health problem, it is recommended that you seek your physician's advice before embarking on any medical program or treatment. All efforts have been made to assure the accuracy of the information contained in this book as of the date of publication. This publisher and the authors disclaim liability for any medical outcomes that may occur as a result of applying the methods suggested in this book.

LIFE IN THE FASTING LANE. Copyright © 2020 by Fung Health Consultants Inc., Eve & Levi LLC, Megan Ramos Nutrition Inc. All rights reserved. Printed in the United States of America. No part of this book may be used or reproduced in any manner whatsoever without written permission except in the case of brief quotations embodied in critical articles and reviews. For information, address HarperCollins Publishers, 195 Broadway, New York, NY 10007.

HarperCollins books may be purchased for educational, business, or sales promotional use. For information, please email the Special Markets Department at SPsales@harpercollins.com.

FIRST EDITION

Designed by Bonni Leon-Berman

Library of Congress Cataloging-in-Publication Data has been applied for.

ISBN 978-0-06-296944-6

20 21 22 23 24   LSC   10 9 8 7 6 5 4 3 2 1

CONTENTS

PART VI: PROBLEM SOLVE YOUR FAST

INTRODUCTION

EVE MAYER

I grew up in south Louisiana, where you don't eat to live. You live to eat! If Willy Wonka had set up shop in my hometown of Thibodaux, he would have specialized in crawfish, gumbo, boudin, and étouffée instead of lollipops, jawbreakers, and gumdrops.

To top it off, my mom is one of the best cooks in the universe, and growing up in our family, we followed the saying *Laissez les bons temps rouler*, meaning, "Let the good times roll." We celebrated absolutely everything, and those festivities—shared with friends, family, and neighbors—centered on food. Cake was love. Crawfish fettuccine was happiness. Fried beignets dusted with generous amounts of powdered sugar meant community.

When I was eight, my mom was diagnosed with a devastating chronic disease for which there was no known cure. For thirty-four years, I watched her fight for her life. She went to specialists all over the country, and she dealt with treatments and medicines that often made her feel even worse. Thankfully, in 2016—when I was forty-two years old—she finally conquered her disease. But until then I wasn't sure if the person at the center of my life would make it another year, and I adopted unhealthy behaviors to cope. I buried my feelings in food: hiding it, sneaking it, and gorging myself multiple times a day. I checked out of my life by letting my brain sail away on a high of waffles, fried chicken, Cajun sausage, and anything sugary I could find in the house. I devised my very own brand of carbohydrate-induced meditation, without the healthy benefits of mental peace.

I've been fat my entire adult life, and at my peak I swelled to a size 26 at three hundred pounds. Every diet plan I tried worked for a short while, but because I always felt hungry, I'd give in, break my diet, and gain back more weight than I had lost. Like many of you who are in a similar boat, I've frequently felt like a failure. I've been ashamed at the doctor's office, at the pool, and in the plus-size store. I've felt embarrassed at the gym, at restaurants, and at family reunions.

In 2018, I decided to try to lose weight again, this time by following a low-carb, high-fat diet. I secretly assumed this diet would fail, too, but, after a month, something felt different. I wasn't hungry every moment like I had always been. After a few months, I'd lost about thirty pounds, but then I started to stagnate. Worried that the weight would pile back on like it always had, I asked for advice from my friend Dr. Suzanne Slonim. She suggested I buy *The Obesity Code* by Dr. Jason Fung.

When I started to read Dr. Fung's book, I was on a plane, buckled in for a four-hour flight. Within minutes, I could *not* put it down. *The Obesity Code* validated my low-carb, high-fat approach to eating, but then Dr. Fung suggested something I hadn't expected. He recommended that people who struggle with their weight benefit from practicing fasting.

What? I had never missed more than one meal in my life unless I was under some sort of medical direction to do so! But the studies Dr. Fung cited in his book made sense to me, and so I decided to give fasting a try. That decision changed my life. I began to lose weight again, I felt healthier than I'd ever been, and my body began to change in ways I never could have imagined. Best of all, the constant hunger messages flooding my brain stopped for good.

That's right. *I wasn't starving all the time.* And when I did feel hungry, it didn't really bother me. I had been worried that I'd pass out if I skipped more than two meals, but I didn't. I thought fasting would make me tired and give me brain fog, but it didn't. I thought not eating would slow down my metabolism, but the opposite happened. I felt like a new woman.

I began to question everything I'd ever learned about losing weight and improving my health, and then I got furious. Where had this information been all my life, and why was I just hearing about it now, after I'd been through *so much*?

I reached out to Dr. Fung, and when we spoke, I knew I'd found a brilliant, kind soul willing to collaborate with me. He also introduced me to his health educator, Megan Ramos, and I felt a connection with her as soon as she explained her own struggles with weight and a host of other medical conditions. Within a month, we formed a plan, and the book you hold in your hands is the product.

In *Life in the Fasting Lane*, we want to empower you to approach weight loss and health in a whole new way. Maybe you've Googled fasting, discussed it with a friend, seen it on the news, or heard one person say it was amazing—and then another claim that it will cause you to starve to death. It seems that the opinions on fasting are as numerous as the stars in the sky, and much of the information can be so complicated and overwhelming that it makes you want to give up before you begin. You may be under the impression that fasting is only for those struggling with obesity, like I was. It's not; fasting can help you lose five or ten pounds—or more or less, depending on your goals. Maybe you need an approach to eating that goes beyond weight loss. Can fasting help sharpen your mind and reduce your risk of cancer? It sure can. Are you desperate to improve your polycystic ovary syndrome, type 2 diabetes, fatty liver, heart disease, and more? Fasting can help.

You need a friend who will tell you the absolute unfiltered truth about fasting, and, in this book, you have three: a veteran of the diet wars (me), a top fasting researcher who has endured her own health struggles (Megan Ramos), and a pioneering doctor (Dr. Jason Fung). Together, we have been there and done that, and we can give you answers about fasting without the sugar coating.

This book is more than just a step-by-step fasting plan. At its core it's a lifestyle guide that will help you prepare yourself, your kitchen, and your family for your new eating routine, as well as troubleshoot common concerns surrounding fasting, like how to deal with holidays and vacations, and how to handle any unexpected side effects. I'll spell out exactly how to prepare your mind to fast, help you find your fasting lane, and give you a plan for sustaining the new, healthier you. I'll tell you exactly why you do not deserve the blame for your weight gain and why this time it *will* be different. I'll hold your hand on this new, exciting journey, and, when it's all done, we'll celebrate your success.

You have questions, and the three of us have answers. The doctor, the layperson, and the researcher: this is the team you need on your side. We've got your back, so let's go!

MEGAN RAMOS

Nearly a decade ago, I suffered from polycystic ovary syndrome (PCOS), nonalcoholic fatty liver disease (NAFLD), and type 2 diabetes. I was also overweight. Today, I am disease-free, I've lost more than eighty pounds, and I'm living out my good health through my career. I'm a clinical researcher who focuses on preventive medicine, and I educate people about how fasting and proper nutrition can help them lose weight and improve their overall health.

For the first twenty-seven years of my life, I ate whatever I wanted and didn't gain an ounce. I was that obnoxious girl walking around in size 0 jeans with a soda in one hand and a bag of chips in the other, and in one of my high school yearbooks, my best friend wrote, "I hate you because you can eat all the chicken nuggets and fries you want, and still seem to lose weight." While I was definitely skinny, I wasn't healthy—in my mind or in my body. In fact, I was constantly fooling myself, thinking that weight was an indicator of physical well-being. But evidence of the real truth lay in the diseases that had caught up with me in middle school.

When I was twelve years old, I was diagnosed with nonalcoholic fatty liver disease, a condition in which excess fat builds up in liver cells. Then, when I was fourteen, I found out I had PCOS, a disorder characterized by cysts throughout the ovaries that leads to irregular or absent ovulation. I was so thin that my doctors didn't advise me to change my diet, but instead assumed that I would grow out of these conditions. They were wrong. Nothing got better with time. I got worse as I continued down an unhealthy path, stuffing myself with junk food without understanding the consequences. Was I using food to cope, like Eve did? Possibly. After all, my beloved mother was sick, too.

My mom had suffered from several metabolic and genetic conditions during most of my childhood, and she visited doctor after doctor and endured countless surgeries over the years. One of my most vivid memories is hearing her scream in pain in the hallway of an emergency room, waiting to be admitted. I decided no one should be sick like that—much less see their mom suffer so terribly—so I vowed to become a doctor when I grew up. I wanted to be someone who held the possibility of making you better—just like that. At age fifteen, I got a summer job in medical research at a private clinic with a group of nephrologists, or kidney disease specialists, including Dr. Jason Fung. In my work with them, I met many beautiful people with type 2 diabetes who were developing kidney failure from their disease. My research focused on finding ways to detect this kidney damage earlier because, if we could do that, we could potentially prevent full-scale failure. I worked with these doctors throughout my high school and college years and loved every minute. But, eventually, I hit a wall. I realized it didn't matter when we caught the kidney disease; most of the time, it would just keep progressing. Early diagnosis felt worse than living in ignorance, and I remember thinking how terrible it must be to live your life knowing what might kill you.

Yet I had chronic diseases, too, and I was doing *nothing* about them. Worse, I was telling myself I was passionate about preventive medicine, but I was slowly killing myself with food. I guzzled diet sodas at 5:00 a.m. and snacked on sugary treats all day long. I downed bags of junk food while my former partner ran errands. I'm fairly certain that I was a food addict, and while I knew that the case of diet soda I stashed inside my car and the bag of pretzels I hid in my purse weren't healthy, I just couldn't help myself.

Everybody has a vice, and mine was food—not cigarettes, drugs, or alcohol—so I justified it as safe. Food is sold everywhere, legally, and carbs were the food group the government and my doctors told me to eat. They were what I was served at school and at home, by my parents. How could they be *that* bad? And, most of all, I was so skinny, so wasn't I doing something right?

I wasn't. My PCOS became so severe that, at the age of twenty-two, my doctor suggested I was probably infertile. For my whole life, I'd wanted nothing more than to become a mother and now that dream might never come true.

Five years later, when I was diagnosed with type 2 diabetes, it was the worst day of my life—even more terrible than when I found out I was likely infertile. When I heard the news, I remember feeling my heart beat so fast, I thought it was going to explode. Everything felt foggy, and I began gasping for air. This was my first anxiety attack.

I was only twenty-seven, and when my doctor handed me my lab results, it felt as if he had given me a death sentence. What kind of life was I facing? Would I suffer kidney failure at thirty-five, just as my research subjects had? Would I get Alzheimer's at forty? Or how about a heart attack at forty-five, followed by a stroke at fifty?

I went home, threw myself onto my bed, and burst into tears. Forget helping people through medicine. I would give up my dream of being a doctor and do something else.

When I finally settled down, I decided I needed to do whatever it took to reclaim my health. My first step was to start eating regular, healthy meals. As a Canadian, I turned to Canada's Food Guide (similar to the guidance provided by the USDA in the United States). I figured, since experts created it, I'd follow its advice to a T. And I did, eating three meals a day, plus several snacks in between. Guess what happened? Instead of being a skinny sack of fat, I became a large sack of fat.

Months later, desperate for a solution, I thought of Dr. Fung—with whom I still worked—and suddenly knew that my diabetes diagnosis might be the greatest blessing of my life.

Always someone to think outside the box, Jason had just started doing research into fasting. One afternoon, I heard him talking to a small group of people about how fasting might help reverse type 2 diabetes, and I thought, *No way. This is too extreme.* But I had nothing to lose. In fact, I had everything to gain.

I spoke to Jason and immediately started fasting and applying his principles of good nutrition, eating a variety of low-carb, healthy-fat, whole foods. Within weeks, I realized that, for my whole life, just about everything I'd learned about nutrition was wrong. It's been eight years since I began following Dr. Fung's recommendations, and I've maintained an eighty-six-pound weight loss. I have completely reversed my type 2 diabetes, fatty liver, and PCOS.

Today, I'm a very happy and healthy thirty-five-year-old, who is privileged every day to help clients successfully lose weight and reverse their type 2 diabetes. I'm still working with Jason, and, together, we created a program called The Fasting Method. It's a Toronto-based online community staffed by fasting coaches who partner with clients to help them lose weight and improve their chronic health conditions. I also consult with clients one on one, giving them the kind of lifesaving advice my doctors never gave me. This is one of the most satisfying parts of my job because I get to know women like Jennifer.

Like me, Jennifer was overweight and had PCOS. The condition also caused her to have acne and male-pattern hair growth, an unfortunate side effect of the hormonal fluctuations the disorder causes. She didn't get her period until she was eighteen, and after that, she menstruated at most once a year. After years of failed attempts to get pregnant, she and her husband decided that they would try—at most—three rounds of IVF. Just to be safe, they would also fill out adoption paperwork and hope that— one way or the other—they would have a baby.

Hundreds of hormone shots later, through all three rounds of IVF, Jennifer's follicles never matured, so doctors weren't even able to extract—much less fertilize—one of her eggs. Thankfully, an adoption came through, and Jennifer and her husband welcomed a beautiful baby boy named Nico into their family.

Jennifer still worried about her health and weight, though, so she consulted with one of our fasting coaches, who guided her through a program of sugar reduction, low-carb eating, and fasting. Jennifer lost some weight, and her menstrual cycle began again. On a whim, she decided to try a fourth round of IVF. She got pregnant, and her second son, Oscar, was born when Nico was two and a half. She continued her healthy habits, and, three years later, she spontaneously became pregnant with her third son. Jennifer is now a healthy, slim mother of three with regular periods and a happier life than she could ever imagine.

Because of fasting, I have confidence that, like Jennifer, I'll be a mom someday. Until then, I am excited to continue the work I am doing to help people change their lives for the better. Together with Dr. Fung and Eve Mayer, I'm here to be your guide, offering a researcher's perspective on how fasting can help you lose weight and stop chronic health conditions in their tracks.

DR. JASON FUNG

I am a nephrologist—or kidney disease specialist—who completed medical school and internal medicine residency at the University of Toronto, then finished my fellowship at the University of California, Los Angeles. Over the last twenty years, I've worked day in and day out to treat my clients' kidneys, supporting these vital organs' normal functioning. I've prescribed the proper medications, recommended the right treatments and surgeries, and followed the correct procedures to help those with renal issues, including stones, diabetes, cancer, inflammation, and more. So, it always seems a little strange to me that I am now practicing obesity medicine, trying my best to take people *off* their medications, escape the surgeon's knife, and avoid dialysis. Basically, my life's mission is to put nephrologists like me out of a job.

Why? Because a decade ago I noticed a disturbing pattern. In the past, the most common cause of kidney disease was high blood pressure, followed by type 2 diabetes. Over time, as proper screenings and the introduction of blood pressure medications helped reduce diseases caused by hypertension, type 2 diabetes surpassed it as the main cause of kidney disease. Medications and technology were obviously not helping these people, and I became increasingly aware that my efforts to treat kidney disease with drugs, dialysis, and more were never going to be successful on a large scale because they were not addressing the root cause of the problem. It was clear that excess body weight, which leads to type 2 diabetes, was the true culprit. Therefore, the only logical solution is to help people *lose* that excess weight.

But how can effective, long-term weight loss be achieved? How can people best reach their weight-loss goals and improve their health? For decades, the prevailing wisdom from doctors has been to eat less and move more. But that doesn't work for the vast majority of people, and countless scientific studies (which I will cite in this book) have proved that calorie restriction is ineffective. Everybody—and I mean *everybody*, myself included—has tried this diet and failed, whether they wanted to lose 5 pounds or 205. Unfortunately, I learned next to nothing about nutrition and weight loss in medical school, so I made it my job to understand both. Weight was arguably the most important determinant of my clients' health, so I knew I had to become an expert on this topic.

But learning new material isn't nearly as difficult as un-learning the failed paradigms ingrained in my mind, and most of what I thought I knew about weight loss—or learned in medical school—has proved to be completely wrong. Caloric restriction is a case in point. In medical school, we were taught that losing weight is a simple matter of eating fewer calories than you expend. "Calories In, Calories Out," right? The truth is that this strategy will *not* help you lose weight, and that's not just my opinion. The success rate of calorie restriction is roughly 1 percent. Obesity has become a global epidemic, even as people have counted their calories more obsessively than ever.

Given the importance of weight loss to health, and particularly in kidney disease, I reviewed the scientific basis of this advice. It stunned me to discover that this entire theory of calorie restriction is without scientific merit. There are no physiologic pathways in the body that rely upon calories. There are no studies that prove that reducing calories reduces weight. On the contrary, every study shows that calorie restriction is futile. If we already knew it was pointless, then why were medical professionals championing this failed method? It boggled my mind.

I decided to look for more successful methods of weight loss, and I found some time-tested strategies that had been forgotten. Soon, in addition to recommending that my clients eat less sugar and fewer refined grains, I introduced them to fasting. This advice was transformational. These people lost weight, adopted healthy habits, and improved many of their chronic conditions.

But there's a component of fasting that extends far beyond what you see on the scale or what shows up in a blood test: the mental and emotional issues surrounding weight loss and poor health, including addiction, shame, and guilt. It is as imperative to address these struggles as it is to tackle medical issues.

That said, I recognize that I'm probably not the best person to address the mental and emotional side of weight loss. I've pretty much stayed the same weight since high school, and I actually wore a pair of pants I'd had for thirty years until fairly recently, when my embarrassed wife threw them out. Sure, I've gained a few pounds now and then, usually after a holiday or vacation, and I lose a little, usually when I get really busy. So, while I understand the struggles of weight loss, I don't relate to them in any serious personal sense.

However, I'm certain that Eve Mayer, who is brilliant and articulate, can put a human voice to these issues, and my longtime colleague Megan Ramos knows obesity from both a personal and professional perspective. Together, we hope to show you how a lifestyle of fasting can help reverse weight gain *and* a host of chronic conditions. This is why we are writing *Life in the Fasting Lane*. To teach. To learn. To laugh. To cry. To form community. To break down myths and stigmas. Mostly, we're writing this book together to help you understand a beast that we are *all* trying to tame: obesity.

PART I

Fasting, Food, and Hormones

CHAPTER 1

The Science
of Fasting

EVE MAYER

Science. It's an intimidating word to me, and always has been. Even in high school, where I steadily earned As and Bs without much effort, I was haunted by the one D on my report card in honors chemistry.

This fear and inability to understand basic science helped keep me fat for twenty-four years. Today, I'm no longer obese, and science doesn't terrify me. Have I suddenly become a scientific savant? Heck, no! Sometimes I still get confused by the simplest terms. But *something* has changed, and I know that part of the reason I shed so many pounds and got healthier was that I became curious about what was going on in my body.

I'm going to talk to you about the science of fasting in the way I wish someone would have spoken to me: in very simple terms. Dr. Fung and Megan will continue with deeper, more detailed explanations about what happens in your body after you eat, but allow me to describe my own personal experience—and how learning about metabolism, digestion, hormones, and more has helped change my life.

For years, I was fat, and I didn't want to be. I also struggled with prediabetes, infertility, allergy issues, sinus infections, joint pain, bronchitis, and pneumonia. I went to one doctor after another to solve my problems, and because I'm a hard worker, I did what they said. I ate fewer calories. I worked out. I took medication. I went to therapy. I ate more fruit and vegetables. To get my slow metabolism to speed up, I consumed smaller meals more often. I had lap-band surgery twice, and later I had the size of my stomach permanently reduced with the gastric sleeve procedure. Results varied with each of these actions. But, most often, I lost weight, only to have it return later. Through it all, I never uncovered the root of the problem.

I thought my body was broken. Then I tried a new approach.

At the beginning of 2018, I cut sugar out of my diet and greatly lowered my carb intake, and an astounding thing happened. I was no longer hungry all the time. It was a dramatic—and more than welcome—change, and suddenly I wanted to know why and how my new approach had worked. When I read *The Obesity Code*, a light came on in my head, and I realized:

Being fat is a problem with my hormones. What I eat and when I eat affects my hormones. Therefore, if I change those two things, I can lose weight.

Doctors had told me I was prediabetic and that there were issues with my insulin, but they had never explained to me what the heck that meant. Why is insulin important? What does insulin do in the body? What is insulin resistance? Why do diabetics take more insulin if increasing insulin is a problem? Why am I taking metformin other than the fact that it will help me not go from prediabetic to diabetic?

Dr. Fung's approach explained everything to me. Now I understand that my body can either be focused on storing energy or burning energy. But not both at the same time. When I eat often, my body becomes busy packing away energy as fat. When I eat *less* often, my body has more time to burn energy—and fat. Fasting allows my body to focus its efforts on *using* energy instead of *storing* energy. I still have warehoused energy in my body in the form of excess fat. My metabolic and digestive systems are completely capable of using that fat as energy—but not unless I give them the opportunity by not eating for a period of time.

I believe this science because of how I feel. My health issues are gone. I am not prediabetic, I rarely get sick, I take no daily medications, and I feel like a million bucks. When I eat often, I feel hungry, tired, and down—and this is because of my hormones, not because I ate a certain amount of food. Fasting is sort of like a good night's sleep. I go to bed to let my body and mind rest. Overnight, my body can focus on replenishing and repairing rather than the million things it has to do when I'm awake. Sleeping also allows my mind to process everything that has happened and sort out what to do with all that information. It's an efficient period of bodily rejuvenation—just like fasting.

If you are fascinated yet overwhelmed by the science of fasting—like I was—take it easy and be kind to yourself. Fasting is worth exploring whether you want to lose three pounds or three hundred pounds, or simply improve your health. The best scientific proof you can ever get is trying it and feeling the changes in your own body.

JASON FUNG

There are so many reasons to make fasting a part of your life. From a purely medical standpoint, many diseases are caused in part by excess body fat. Being overweight increases your risk of heart disease, stroke, and cancer. Losing weight increases your high-density lipoprotein (HDL) or "good" cholesterol levels and lowers your triglyceride levels, which helps reduce the risk of those same diseases. Excess weight may raise your blood pressure, lead to or aggravate arthritis, disrupt your sleep, cause back pain, cause liver disease, and more. Type 2 diabetes, which is closely related to increased body fat, is also the number one cause of blindness, kidney disease, nontraumatic amputations, and infections. As a nephrologist, I've seen forty-year-olds with type 2 diabetes go into renal failure and need dialysis, which is a life-sustaining treatment they endure for the rest of their years. I've seen fifty-year-olds with type 2 diabetes develop poor circulation in their legs and require amputation. I've witnessed more people with type 2 diabetes lose their vision than I can count. Yes, in those cases, losing weight would have greatly improved my clients' health and helped them avoid the diseases and side effects that diminished their quality of life—or, tragically, ended their lives entirely.

But, as Megan discussed in her introduction, there are many people who are not overweight—as defined by body mass index (BMI)—and are still metabolically unhealthy. And there are many people who are overweight—again, defined by BMI—and are metabolically healthy. So, while weight is not the entire story, fasting has been shown to help decrease the prevalence of many metabolic syndromes, including type 2 diabetes.

I know this may be overwhelming, or worse, too good to be true. How can skipping a few meals—or even just one meal a day—

make such a difference in your health? Our client Natasha offers an example of just how beneficial a fasting lifestyle can be.

Natasha was diagnosed with type 2 diabetes in early 2012. Even though she'd tried altering her diet, exercising, and taking metformin (a drug prescribed to treat diabetes), almost nothing worked. She didn't lose weight from her small, 5-foot frame; metformin made her miserable; and her blood sugar spiked even when she ate a tiny amount of carbs.

Natasha had tried fasting, and she liked it, but she was afraid to fast longer than a day. Her fasting coach eased her fears about extended fasting, and, today, Natasha refrains from eating for forty-two hours two or three times a week. Her blood sugar level is now within a normal to prediabetic range, and she's down to a size 2. She looks great. She feels great. And, best of all, her health has been restored. Fasting has changed her life.

I know there are those of you out there who are scared to fast for a few hours. Even the idea of cutting out something as simple as snacks may make you feel anxious. But, if you're like Natasha, it could just be fear that's holding you back. Or, perhaps, you don't understand the science behind why fasting works, and what it can do for you. Knowledge is power, so allow me to explain how the foods you eat impact your body, why they may lead to hormonal fluctuations that cause weight gain and chronic disease, and how fasting can help.

Digestion, Hormones, and How Food Is Stored for Energy

The second that food enters your mouth, your body begins the hard work of turning that food into cellular energy. However, the path isn't always easy or straightforward, and if you eat the wrong foods or consume them in excess, your body may develop problems.

The body's endocrine system includes a vast network of glands that release hormones into the bloodstream to regulate all the body's functions, including sleep, metabolism (the conversion of food to energy for cell function), reproduction and sex drive, mood, hunger, and more. When we eat, the pancreas—a narrow, six-inch-long organ that sits behind the stomach and is part of both the endocrine and digestive systems—secretes the hormone insulin. Insulin signals to the rest of the body that food is now available to process into energy, and this food energy (calories) needs to be stored away for the future.

The body stores food energy in two different ways: as sugar and as body fat. Sugar is available for quick energy, while fat is kept in reserve, available to burn when our body doesn't have blood sugar at the ready. Let's talk about sugar first, as the stable regulation of blood sugar—also known as glucose—is one of the central benefits of fasting.

One of the easiest ways to spike your blood sugar is to eat carbohydrates, which, chemically speaking, are chains of sugars. When we eat carbohydrates, some of this sugar is used by cells in the kidneys, liver, brain, and more. If there are carbs left over, they're stored in the liver as glycogen, another chain of sugar. We'll come back to glycogen in just a minute.

The other way our body stores energy is body fat. When we eat dietary fat (found in all kinds of plant and animal foods, from potato chips to red meat to milk), the individual fat molecules, called triglycerides, are absorbed directly into the bloodstream and delivered to fat cells. If we eat too much glucose and exceed the body's ability to store it in the liver as glycogen, the liver converts this glucose to triglycerides. The triglycerides then feed fat cells.

These two systems of energy storage—glycogen and fat—are complementary. Glycogen is easy to use and simple for the liver to process, but the liver has limited space to store it. Body fat is harder to get to and more difficult for the liver to break down, but it offers the advantage of unlimited storage space (as anyone worried about the rolls of fat on their belly knows all too well!). Think of glycogen as a refrigerator. You can easily put food into it and take it out at a moment's notice, but you only have so many shelves. Think of body fat like a basement deep freezer. It's harder to get to, harder to cook the food in it (because it's frozen), but it's huge and almost never full.

Insulin and the Development of Diabetes

As I stated previously, insulin is the hormone that signals your body when it's time to convert food into energy. But its job doesn't stop there. Insulin also regulates the body's glucose levels, making sure they don't spike or plummet. It does so by helping to extract glucose from the blood to store it in the liver as glycogen or in the body as fat. Because the body needs fat for protection, warmth, and energy in times of famine, insulin also prevents us from using too much body fat as a source of energy.

If your insulin levels are high, the body will put food energy into storage, both in the fridge and the freezer. Problems begin, however, when your pancreas goes into overdrive, secreting too much insulin. How does this happen? All foods, which contain a variety of macronutrients (protein, fat, and carbohydrates), will stimulate insulin production to some degree, but certain foods are more effective than others. The worst offenders in this regard are refined carbohydrates, like white bread, sugary drinks, cakes, and cookies.

If we eat lots of sugar or carbohydrate-rich foods too often, as is the case with the typical Western diet, where people regularly eat six or seven carb-heavy meals or snacks a day, our insulin levels will spike. High levels of insulin tell the body to keep trying to store food energy, preventing us from burning our fat stores. We are, in essence, continuing to restock the refrigerator, while wondering why our basement freezer is bursting at the seams.

Eventually, when there is too much insulin flooding your system, the cells in your pancreas that produce it can no longer respond, and your blood glucose levels become high. If they *stay* high, you can now call yourself one of the estimated 500 million people in the world with type 2 diabetes.

Measuring and Treating Diabetes

If you become diabetic, your symptoms may include increased thirst, fatigue, blurred vision, hunger even though you're eating more than you normally do, frequent urination, tingling, pain, or numbness in your hands or feet, or cuts or bruises that are slow to heal. But you may not have *any* symptoms. Many people discover they're at risk of diabetes or already have it only after they have a blood test.

There are several tests doctors use to determine if a person has diabetes, but I'll talk about two of them because many of my clients have them regularly—and many see their results improve dramatically after they try fasting.

The first test is the A1c test. This simple blood test measures what percentage of your hemoglobin—a protein in red blood cells that carries oxygen—is covered with sugar. A1c measures average blood sugar levels over two to three months, so one carb-heavy meal won't necessarily impact the results. People without diabetes have low A1c levels, between 4 percent and 5.6 percent. If your A1c levels are between 5.7 percent and 6.4 percent, you're at risk for developing diabetes, often referred to as prediabetic. And if your levels are over 6.5 percent, you have type 2 diabetes.

The other test is called the fasting plasma glucose test, or FPG. This test measures blood glucose levels at one point in time, and it's given after you've fasted for eight hours, typically in the morning. A high result—indicating you have diabetes—is anything over a level of 126 mg/dL. A prediabetic level is between 100 and 125 mg/dL, and anything under 100 mg/dL is considered normal.

If your test results are in the prediabetic range, you'll need to adjust the foods you eat and perhaps consider some of the medications I'll discuss below. But remember to focus your health efforts not just on diabetes. Midrange A1c or fasting glucose levels also mean you're at risk for heart disease, stroke, cognitive difficulties, or insulin resistance (a disease in which your body doesn't respond well to insulin and raises your blood sugar).

In addition to weight loss, exercise, and diet modifications—typically a diet low in sugar and carbs—the most common treatment for diabetes is prescription medication. Metformin is the gateway drug for diabetes treatment, and it works by limiting the amount of glycogen your liver converts into glucose, as well as helping your body use insulin more productively. Other drugs—including sulphonylureas—help the body produce more insulin or become more sensitive to it, excrete glucose into the urine, or slow digestion. Typically, the last resort for diabetes treatment is insulin, given by subcutaneous injection.

However, it's disheartening—to say the least—that fasting is not recommended by the health community. Why? Because more than any drug or diet modification, fasting helps control your insulin. Type 2 diabetes is essentially a disease of too much sugar and too much insulin. What decreases sugar and insulin? Fasting. When your insulin is in check, your blood sugar stays in check, your weight stabilizes or decreases, and your risk of developing any number of chronic health conditions goes down.

Where Fasting Comes In

If I could sum up fasting in one sentence, I would say this: fasting regulates your hormones. It's more than a diet; it resets your body's internal controls, allowing it to burn the right amount of energy to keep you alive.

When we don't eat (fast), insulin levels fall, and this signals our body that no more food is available. In order to survive, the cells then draw upon the stored energy supply, either in the form of glycogen or, if that's been fully expended, fat. This is the reason we don't die in our sleep every night, or why we can live a few hours—or a few days or more—without eating. The body has a wondrous ability to store food energy, then find it in either the refrigerator or freezer to burn it.

Therefore, it figures that the most logical solution to keeping our blood sugar levels stable, allowing the body to continue to use its stored reserves of energy, is to fast. By not eating, we allow insulin levels to drop, which tells the body that food is no longer available and that it's time to eat some of the food in the fridge (glycogen) or freezer (body fat). Weight loss and preventing type 2 diabetes—as well as a host of chronic conditions I'll outline in the next chapter—are about correcting the underlying hormonal imbalance that caused obesity. This hormonal imbalance, again, is an insulin level that stays high for too long.

Fasting and Metabolism

But what about metabolism? Doesn't fasting kill it, as many of you have heard? For that matter, what *is* metabolism? Our metabolism, or basal metabolic rate (BMR), is the amount of energy (calories) required to keep our bodies alive while we are at rest. BMR is the measure of what we need to keep our body's very basic functions—such as brain activity, circulation, and digestion—chugging along. If you have a high metabolism, your body burns energy more efficiently, and you tend not to put on weight rapidly. If it's lower, weight loss will be more of a struggle.

Our BMR is not fixed. Our bodies may increase or decrease BMR by 30 to 40 percent depending on our diets, level of activity, age, body temperature, and more. But, from a dietary perspective, the most significant determinant of BMR is insulin.

The body only exists in one of two states: the "fed" state, after we've eaten, and the "fasted" state, when we have not eaten. In the fed state, insulin levels are high, and the body wants to store food energy as sugar or fat. Our metabolism is humming. In the fasted state, when insulin levels are low, the body wants to burn stored food energy. So, we're either storing calories or burning calories, but not both at the same time.

If we elevate insulin levels (by eating foods that stimulate insulin) and keep them persistently high (by eating constantly—say, by consuming six or seven snacks or meals per day instead of three), then the body must stay in the "fed" state. The body stores calories because those are the instructions we've given it. If all the calories are going into storage, then there are fewer calories to use, and therefore the body must slow down its energy expenditure, or BMR.

Suppose we are eating 2,000 calories per day and burning 2,000 calories per day. We neither gain nor lose body fat. We now reduce our calories to 1,500 by eating high-carb, low-fat foods six or seven times per day, as many health professionals urge us to do. Insulin levels stay high but calories drop. Now, the body cannot burn body fat stores because insulin is high, and we are in the "fed" state. With only 1,500 calories coming in, the body must reduce its calorie expenditure to 1,500 as well. We cannot make up this caloric deficit because insulin prevents us from burning fat. We are in "fat storage" mode. This is the dirty little secret of the low-fat diet. At first, the weight comes off, but as our BMR drops, the weight plateaus and then eventually returns.

What happens during fasting? A study of four consecutive days of fasting—that is, four full days without any food to eat—showed that BMR increases by about 10 percent. Yes, metabolic rate increases when you don't eat. Why? We know that fasting decreases insulin but increases counter-regulatory hormones, so called because they run counter to insulin. If insulin falls, these hormones go up. If insulin rises, these hormones go down. The counter-regulatory hormones include noradrenaline (responsible for stimulating muscle contraction and heart rate), growth hormone (stimulates cell growth and regeneration), and cortisol (the so-called stress hormone, responsible for triggering motivation and action). If noradrenaline increases, then metabolic rate is expected to also go up.

The increase in BMR is likely a survival response. Imagine that you are a caveman. It's winter, and there is nothing to eat.

If your metabolic rate decreases, that means that for every day you do not eat, you get a little weaker. This makes it that much harder to find and hunt for food. It's a vicious death spiral. As you get weaker, you'll be less likely to find food. As you don't find food, you get weaker. If this is what happens to your body, you would not have survived. Your body is just not that stupid.

Instead, your body switches fuel sources. Instead of relying on food, you turn to stored food (body fat), and your body does not shut down. It ramps up by increasing noradrenaline, cortisol, and the other counter-regulatory hormones. You power up by using a different fuel source. Concentration increases. Focus increases. So BMR increases during fasting. If you maintain BMR during weight loss, as opposed to burning 500 calories less per day, that is a huge advantage.

So, the key to the energy balance equation of "Calories In, Calories Out" is not the number of calories we eat and the exercise we do. That is virtually irrelevant. The key is to control hunger and maintain basal metabolic rate. In order to do that, we must eat foods that increase satiety hormones and keep insulin (fat-storing hormone) low. Fasting provides the hormonal changes necessary to successfully lose weight in the long term. Hunger decreases while BMR is maintained. And guess what? Fasting has been used for thousands of years, during which time obesity has been no more than a passing footnote in the pantheon of human illness.

MEGAN RAMOS

When I think about all the clients I've met over the years, I can't even count the number of conditions that fasting has helped improve. Diabetes and obesity are the two obvious ones, but I also think about a woman named Marta, who suffered from acne, PCOS, joint pain, asthma, allergies, gall bladder disease, restless leg syndrome, mood swings, reactive hypoglycemia, fatigue, heartburn, Hashimoto's, and sleep apnea—in addition to being overweight and receiving a devastating diagnosis of type 2 diabetes. Until she saw success with fasting, Marta didn't know that all of these conditions can be related—or that fasting could fix them, as it did for her.

But what about the more serious illnesses, such as cancer or Alzheimer's disease, that Marta might have developed in the future if she kept up her unhealthy lifestyle? There are now credible and compelling scientific studies showing that the science behind fasting doesn't end with obesity, diabetes, and blood sugar regulation. Fasting is a lifestyle that may *prevent* a host of chronic conditions that seem to have nothing to do with the food we put in our mouths. Fasting can have significant benefits for your brain, your mood, your risk of cancer, and more.

Fasting and Your Brain

The brain is a remarkable, complex, and resilient organ, and it's one that's not impacted negatively by fasting. So, if you're concerned that fasting will cause you to be mentally slow, dull, or foggy, worry no more.

Fasting may even *help* your brain. I say "may" because unfortunately, no authoritative studies exist on fasting's impact on the brain. However, two human studies—one that measured brain activity after a twenty-four-hour fast and one that measured it after two days—established that reaction time, memory, mood, and general function were not impaired by fasting. And in a study of rats who were put on a fast, the mammals improved their scores of motor coordination, cognition, learning, and memory. In addition, they showed increased brain connectivity and new neuron growth. Now, I know rats aren't humans, but these results echo what so many of my clients say: that fasting makes them feel sharper.

Evolution also provides some clues to how fasting can help your brain. During times of severe caloric restriction, the organs of many mammals shrink in order to survive. But there are two exceptions: the brain and the male testicles. Obviously, the testicles stay the same size so that the males of the species can continue to attempt to mate, but what about the brain? Think about how you'd feel if you were starving. You'd want to be sharp and focused so you could search for food, right? That's what happens with most mammals. Conversely, when we eat too much, we may experience brain fog, or what's often known as "food coma." Consider how you feel after a huge Thanksgiving dinner: lethargic, dull, and able to focus only on the thought of a nap.

The most encouraging research I've seen are the animal studies that demonstrate that rats who were subjected to fasts showed fewer symptoms in models of Alzheimer's, Huntington's, and Parkinson's disease. Fasting induces autophagy—a cellular process that helps the body clear out old or damaged cell parts—and, in these studies, rats on a fast saw a decrease in the accumulated proteins that are a hallmark of Alzheimer's. Imagine if fasting could prevent, treat, or even reverse these heartbreaking degenerative neurological conditions? Lives could be saved, suffering reduced, and we'd save tens of billions of dollars in healthcare costs.

Fasting and Cancer

Cancer is the second-leading cause of death worldwide, killing about 10 million people every year. One out of six people will die from it. Many cancers develop due to genetic factors, unintended toxic exposure, viruses, or some other, often unknown cause. For the most part, these unfortunate cases are difficult to prevent. But there are promising studies that show that cancers that were previously thought to be unavoidable may be preventable, in part, through fasting.

One of the keys to these findings, as with type 2 diabetes and obesity, is insulin. If you extract breast cancer cells from tissue, it's quite simple to grow them in a lab. If you add glucose, epidermal growth factor (EGF), and insulin, they multiply rapidly. If you then take away the insulin, they die. Let me repeat that: breast cancer cells proliferate with high levels of insulin and die without it. What lowers insulin levels? Fasting.

The second reversible link to cancer is obesity. A 2003 study released by the American Cancer Society highlighted the findings from 900,000 US men and women. From 1982 to 1998, these people were tracked every few years to determine who had died and how they'd died. At each interval, their BMI (body mass index) was also factored. While all were free from cancer at the start of the study, after sixteen years, just over 57,000 of them were dead from cancer. Shockingly, for those with a BMI over 40, the death rates from all cancers combined were 52 percent higher for men and 62 percent higher for women. BMI was positively associated with death from esophageal, colon, rectal, liver, gallbladder, pancreatic, kidney, non-Hodgkin's lymphoma, multiple myeloma, breast, stomach, prostate, cervical, uterine, and ovarian cancers. Researchers concluded that being overweight or obese accounts for 14 percent of all deaths from cancer in men and 20 percent for women. The evidence was clear: obesity is a major risk factor for cancer. What helps you lose weight? Fasting.

Finally, autophagy may slow down cancerous growths or prevent cancer from occurring—a finding that shocked scientists, who had previously believed that autophagy *increased* cancer growth. A 2019 study published in *Nature* concluded that autophagy played a major role in killing certain cells linked to cancer. When autophagy is stopped, these harmful cells can continue to replicate, fueling the growth of cancer. What causes autophagy? Once again, fasting.

Fasting and Metabolic Syndrome

Metabolic syndrome, also called Syndrome X, is a group of conditions that meet three of the following five criteria: abdominal obesity (as measured by waist circumference), hyperglycemia (type 2 diabetes), high triglycerides, low HDL, and hypertension.

The common factor among these conditions is that they all involve an excess of insulin. When insulin is too high for too long, the body stores more body fat than it needs to. Cells become overloaded with glucose, and they become resistant to insulin. Glucose from the blood can no longer go into the cells, and blood glucose levels become elevated. This is the disease known as type 2 diabetes. When the liver is overloaded with glucose, excess sugar gets stored as fat, and fatty liver develops. Attempting to unload all this extra fat, the liver exports the glucose into the blood, which causes blood triglyceride levels to increase and HDL levels to decrease. In short, excess insulin causes a series of problems that collapse, one by one, like dominoes.

Since metabolic syndrome is a disease of too much insulin, lowering insulin levels is critical to reversing it. Refined carbohydrates cause the greatest increase in insulin, so eating a diet low in refined carbs and sugar is a great start. Because all foods contain a mix of protein, carbohydrate, and fat, eating even those foods that are healthy will raise your insulin level somewhat. This is why fasting is so effective for treating metabolic syndrome. When you refrain from eating, your insulin levels drop and remain at a lower baseline.

Clearly, fasting helps to stabilize blood sugar. But having stable blood-sugar levels is only one of many benefits of a fasting lifestyle. As we will soon see, it can do wonders for the mind as well as the body.

CHAPTER 2

Beyond Science

The Mental and Emotional Benefits of Fasting

MEGAN RAMOS

It's logical to assume that if you're not fat, sick, overmedicated, and exhausted—all situations that fasting can alleviate and reverse—you're going to be a happier person. This is the case with thousands of clients I've seen over the years. As they begin to lose weight, take fewer medications, and suffer fewer of their painful symptoms, their moods improve. They're not depressed anymore. They don't fight with their spouses as much. They begin doing activities they love.

Even if you only want to lose five, ten, or twenty pounds, or if your health problems are minimal, fasting can still change your life. I think of a sixty-seven-year-old client I had, named Paul. Paul started fasting to support his wife, who was extremely overweight and had just received a diagnosis of borderline type 2 diabetes. Unlike her, Paul wasn't bothered by bad health, and he figured the twenty excess pounds he was carrying around were a natural product of aging. But out of devotion, Paul stopped snacking and skipped meals a few times a week, and, within several months, he lost *all* his excess weight. Not just that, but he felt great—physically *and* emotionally.

The mood-stabilizing properties of fasting aren't merely anecdotal. A 2016 study published in the journal *Frontiers in Nutrition* measured the effects of an eighteen-hour fast on fifty-two women whose mean age was twenty-five. The study looked for changes in mood, irritability, sense of achievement, reward, pride, and control. At the end of eighteen hours, the study concluded that, while the women felt more irritable than they had before the fasting period began, they had, on the whole, a significantly higher sense of reward, achievement, and pride.

These findings are consistent with my clinical observations over the years. Some people new to fasting can experience anxiety, which can be attributed to the hormone noradrenaline that is secreted during a fast. Noradrenaline causes your blood pressure to increase, your heart to beat faster, and your nervous system to become more alert. Together, these effects may be experienced as anxiety. Usually, they don't last longer than two weeks, when the body adapts to the increased level of noradrenaline.

However, most of my female clients—of all ages—don't report feeling irritable on short fasts (meaning intermittent fasts of forty-two hours or less, three times a week) either. Instead, I've noticed they may become more emotional or irritable only on extended fasts of five days or more—and only when they're new to fasting. These women may become emotional because, during an extended fast, you start to lose belly fat in large quantities. Excess fat cells produce excess estrogen, and the loss of those cells releases estrogen out of storage and into the bloodstream. So, for a short period of time, your hormones soar, which has an impact on your emotions.

I've found that people who lose the most water weight—an accumulation of fluid in the tissues and body cavities, often leading to bloating—are the ones most likely to report mood swings and irritability. We believe this is due to the electrolyte loss that's experienced when there's a decrease in water weight. This is only a short-term problem, though, and mood tends to stabilize when the body stops dumping excess weight.

JASON FUNG

Being overweight affects more than just your physical health—it also impacts your mental and emotional health. While I strongly believe that we should accept ourselves and others regardless of weight, as a culture we have a long way to go toward acceptance. The truth is that misinformed attitudes about weight pervade just about every part of our society.

These beliefs create an unconscious bias that generates a lot of discrimination. *A lot.* Many people unwittingly perceive overweight people as lazy, gluttonous, and lacking willpower. This is the direct result of the "Calories In, Calories Out" mode of thinking, endorsed by most health professionals and researchers. Those who believe in the erroneous "energy balance equation" imagine that weight loss is such a simple formula that everybody has the knowledge and ability to lose weight, if only they try. You just expend more energy than you put in your body. Therefore, if you're gaining weight, it's because you lack the willpower to get off the couch, put down your fork and knife, or move your body. You have a character defect, and it doesn't seem to matter that 99 percent of people who try a calorie-restricted diet fail to lose weight over the long term. Or that every single study of this "Eat Less, Move More" approach has failed. *Every single one.*

People who believe in calorie restriction are, in my opinion, misguided. The problem is not with the 70 percent of adults in America who are overweight or obese, but rather, with the dietary advice we've been given. Indeed, since the advent of the Dietary Guidelines for Americans in 1977, consumption data from the United States Department of Agriculture shows that Americans have been doing exactly what they have been told to. Americans have consumed less meat and dairy and replaced their animal fats with vegetable oils. They've eaten more grains, fruits, and vegetables. And what has happened? A tsunami of obesity the likes of which the world has never seen.

Nevertheless, because of this prevailing view that obesity is a personal shortcoming, studies consistently find that obese individuals are viewed as less desirable subordinates, co-workers, and bosses in the workplace. They are perceived as disagreeable, emotionally unstable, lazy, and lacking in self-discipline. This always struck me as slightly odd. Considering the lengths to which most people have gone to try to lose weight, "lack of self-discipline" is one of the least accurate descriptions of most of the overweight people in my practice. Women, as is often the case, are judged more harshly than men. A full 60 percent of overweight women consider that they have been on the receiving end of weight discrimination, compared to 40 percent of men.

The problem with most "Calories In, Calories Out" proponents is that they think far too simplistically about the human body. They believe that obesity is only a first-order problem rather than the more complex issue it really is. Let's start with the "Calories In" part of the equation. Most nutrition "experts" say that this is determined by the foods you eat. This is true. But that is first-order, simplistic thinking. What makes you put that food in your mouth? The answer can be many things. Hunger. Emotions. Stress. Medications. You must deal with the root cause of the problem, not the simple first-order cause.

How about the "Calories Out" part of the equation? Most nutrition "experts" believe this is determined by exercise, or the number of steps you take in a day. This only makes up a minority of the calories you burn in a day. The overwhelming majority of calories are used for metabolism, the energy needed for your brain, heart, lungs, kidneys, liver, and other organs and systems.

In the disease of obesity, what causes "Calories In" to exceed "Calories Out," therefore causing fat accumulation? A simpleton would say "How much food you eat and how much exercise you get." The more nuanced understanding of human physiology suggests that the main problem is hunger and metabolism. Yes, you can decide what you want to eat, but no, you cannot choose to be less hungry. Yes, you can decide to exercise, but no, you cannot resolve that your liver will use more energy. So, if you cannot consciously make a decision about hunger and metabolism—the more important root factors of weight gain—then obesity is not a personal failing. It's not a breakdown of willpower. It's a failure of knowledge.

Let's return to the issue of weight bias. The effect of weight on earning potential is staggering, significantly impacting salary, but the result is different for men and women. For women, the skinnier you are, the more money you'll make, even to a startling seventy pounds below average weight. In fact, women are punished for *any* weight gain, and very thin women earn approximately $22,000 more than an average-weight woman. Very heavy women earn about $19,000 less than the average.

The opposite holds true for men, who make more the heavier

they are—except when they reach the top of their profession. Both men and women hit the proverbial glass ceiling if they're obese, meaning they have a BMI greater than 30. According to a 2009 study, only 4 percent of top male CEOs were obese, compared to 36 percent of the general male population. But 61 percent of top male CEOs were overweight (BMI 25–29.9), indicating a level of tolerance for those who are only modestly heavier than the average. The differences for women were much starker. Only 3 percent of top female CEOs were obese, compared to 38 percent of the general population. But only 22 percent of top female CEOs were overweight, compared to 29 percent of the general population.

These statistics are shocking and are part of what drives my work to destigmatize obesity and give everyone the tools to live the healthiest life possible.

How Fasting Differs from Other Weight-Loss Plans

1. **IT'S SUSTAINABLE.** This isn't a short-term eating plan, where you cut out one food group for a few weeks until you've shed the pounds. It's a long-term, sustainable lifestyle.
2. **IT'S FREE.** There are no special foods or gimmicks to buy into. In fact, fasting will save you money.
3. **IT'S FLEXIBLE.** Just skip snacks, skip one meal, or fast for an entire day. Customize a plan that works for you.

CHAPTER 3

Hormones and the Hunger Bully

EVE MAYER

I used to believe that hunger was a bully. It was bigger, stronger, and meaner than I was, and it was present in my home, at work, at my parents' house, on the street . . . *everywhere*. But unlike childhood bullies, I couldn't just walk away from it or report it to my teacher. I believed there was only one way to make the Hunger Bully go away.

Feed it!

The Hunger Bully insisted I consume mass quantities of unhealthy food in order to satisfy it, and over the years I ate so much that the amount required to feel full increased. I even had my stomach surgically reduced—three times!—and I still never got relief. The Hunger Bully always seemed to show up at inopportune moments, when my full attention should have been focused on what was happening in my life and not what I wanted to go into my mouth. My stomach grumbled loudly at my cousin's graduation, my daughter's kindergarten play, and a giant pitch meeting with a client who might help make me rich.

The Hunger Bully knocked on my brain hundreds of times each day, so I ate in excess, usually six to ten times, trying to make it quiet. Sometimes it wanted sugar, so I'd eat enough to give me a high and make the Hunger Bully calm down. That high got shorter over the years, so I binged on more and more sweets. Soon, my sugar high was replaced by pain from eating too much food and a deep, coma-like sleep that temporarily separated me from my feelings in a sweet release.

I hadn't *always* thought of hunger as a bully. When I was young, I saw it as a natural part of life. But when I got fatter and hungrier in my adult years, I just accepted that I was worse at dealing with hunger than other people. I believed that my willpower was nonexistent, and my body and mind were somehow broken. This seemed odd since I could outwork most people in all other situations in my life. Why was I so powerless when it came to weight and health? It just didn't make sense.

My daughter, Luna, was the one who finally gave me the answer. When Luna was in elementary school, she was the target of bullies. One kid in particular became an ever-present problem, and the bullying got so bad that school administrators had to intervene. Nothing changed, and the bully's behavior continued to escalate. The school took a new approach, and we decided to do so at home as well. I began to work with Luna to see what behaviors *she* could change to lessen the chances of being bullied in the future.

When I talked with Luna and thought hard about the kind of kid she was, I realized one of her most notable characteristics was her desire to fit in—and her willingness to do anything to make it happen. That need to be accepted had caused her to take the bully's taunts to heart. She obsessed about his comments, decided everything he said about her must be true, and, instead of fighting back, became powerless around him. She reasoned that if she could only give the bully what he wanted, the suffering would end. It wouldn't, of course. A bully's goal is to cause distress in order to gain power.

As Luna and I discussed her motivations and reactions to the bully, I realized I needed to learn this lesson, too. Hunger was my bully, and only my actions and reactions could give it power. My belief that my hunger would only stop if I gave it the food it wanted was ludicrous; if I just ignored it, like most bullies, it would disappear.

Let me tell you that hunger is temporary. It will not always go away if you give it enough food. Once I changed my way of eating and only saw the Hunger Bully four or five times a day, I began to simply notice it instead of feeling afraid of it. You don't have to give in to it; simply recognize the constant hunger that comes from unhealthy, long-standing habits, and understand that, with fasting, you're working to turn those around.

Today, when I fast I visualize the process of my body choosing to burn stored fat as energy. I even go so far as to place a Band-Aid on a fatty area of my body (like my thigh) and remind myself that when I fast, I am not depriving myself. Even if I get hungry, I am not starving. I am simply feasting on the fat stored in my inner right thigh, which I have saved up for this very day! You can do the same. If you have excess fat on your body, then you already have the energy you need to make it through a day, three days, a week, or more! Food is not required, and it's the Hunger Bully that's tricking you into thinking that it is.

Seven Things I've Learned About the Hunger Bully

1. **HUNGER IS A HABIT**: Hunger often arises around the times that you'd normally eat. When you reduce the frequency of when you eat, hunger soon goes away.
2. **HUNGER IS FLEXIBLE**: When you start to make better choices about the things you put in your mouth, hunger lessens over time.
3. **HUNGER WILL GO AWAY**: If you don't eat when you become hungry, eventually that hunger will pass.
4. **HUNGER IS NOT STARVATION**: You can and will rely on the ample extra fat supplies in your body to sustain yourself while you fast.
5. **HUNGER HAS DIFFERENT CAUSES**: It can be a message from your brain, your body, or both.
6. **HUNGER DOESN'T ALWAYS NEED TO BE FED**: You don't have to give the Hunger Bully food. If you notice it, you can give it water, other liquids, or ignore it entirely.
7. **HUNGER DOESN'T HAVE TO BE A CAPITALIZED WORD**: With strong mental practices and new habits, you can transform the Hunger Bully into just hunger.

JASON FUNG

Probably the number one concern most people have before starting an intermittent fasting regimen is whether they'll get hungry. The answer is yes, but it won't be as bad as most of them believe. Hunger isn't a problem if you manage, treat, and think about it differently than you have in the past. Hunger also doesn't need to be something you fear. Being able to wrap your brain around it can be key to overcoming it.

Hunger and Hormones

Why do we eat? Because we're hungry. What stops hunger? There are certain hormones that make us feel full. These are called satiety hormones, and they are very powerful. The stomach also contains stretch receptors. If the stomach is stretched beyond its capacity, it will signal satiety and tell the brain to make us stop eating.

People often imagine that we eat just because food is in front of us, like some mindless eating machines. That's far from the truth. Imagine that you have just eaten a huge 20-ounce steak. It was so delicious that, even though you thought you wouldn't be able to finish it, you did. Now you're completely stuffed, and the mere thought of eating more nauseates you. If somebody set down another steak and offered to give it to you for free, could you eat it? Hardly.

Our body releases satiety hormones to tell us when to stop eating. And once these kick in, it's extremely difficult to eat more. This is why there are restaurants that will offer you a free meal if you can eat a 40-ounce steak in one sitting. Trust me, they aren't giving away many free meals.

The main satiety hormones are peptide YY, which responds primarily to protein, and cholecystokinin, which responds primarily to dietary fat. The final hormone associated with hunger is ghrelin, which is appropriately called the "hunger hormone." I'll get to ghrelin in a minute.

The reason that most people are hungry all the time is that they've been conditioned to believe that the only thing that matters as far as weight loss is concerned is to consume fewer calories than they expend. Governments throughout the Western world have pushed for a diet heavy on carbs—the very substances that lead to hunger. Think about it. Imagine you sat down to a carb-based, calorie-reduced breakfast consisting of two slices of white bread with jam. How does this meal impact satiety? Sure, it doesn't allow for many "Calories In," but does it control hunger? No. There is no protein to activate peptide YY. There is no fat to trigger cholecystokinin. There is no bulk to activate stomach stretch receptors. The starches in this meal (which are chains of glucose that are essentially mainlined into the bloodstream) increase insulin. We're left hungry because we have not signaled to the body *not* to be hungry. At 10:30, we find ourselves scrounging around for a low-fat muffin that will, once again, leave us ravenous by noon. At lunchtime, we'll look for a large bowl of low-fat pasta and sauce. Already, instead of eating three larger meals, we're on target to eat six or seven smaller meals. By 2:30, we are starving again, so we grab a low-fat granola bar, and then eat rice for dinner, and then scrounge around the refrigerator for a nighttime snack, because we are *so* hungry.

But if you eat bacon and eggs for breakfast—a meal that's high in dietary fat and protein—do you want to eat again at 10:30? No.

The problem becomes amplified if we are eating, as most people are, processed and refined carbohydrates. When you consume refined carbs, your blood sugar levels skyrocket, telling your pancreas to produce a surge of insulin. The job of insulin is to tell your body to store food energy as sugar (glycogen in the liver) or body fat. That huge spike in insulin immediately diverts most of the incoming food energy (calories) into storage forms (body fat). This leaves relatively little food energy for metabolism. Your muscles, liver, and brain are still crying out for glucose for energy, and you get hungry despite the fact that you've just eaten. It's the worst kind of domino effect if you're looking to maintain or lose weight.

In addition, because these processed foods have had most or all of their fiber removed, they don't take up much space when they hit your belly; therefore they don't activate the stomach's stretch receptors. At snack time, most of your ingested calories of food energy have already been deposited into your fat cells, so it's no wonder that you get hungry fast!

Hunger, Fasting, and Ghrelin

What happens to hunger during fasting? People always assume that hunger will increase until it becomes unmanageable. The truth always surprises people. Hunger tends to *decrease* during fasting. Why? There are two reasons, and the first has to do with the two means by which your body derives food energy.

During fasting, your body switches fuel sources. Instead of relying on blood glucose (derived from food) for energy, your body begins burning body fat (which is stored food energy). That switch is your body entering a state called ketosis. Once you begin ketosis, your body has access to hundreds of thousands of calories stored in the fat. You're feeding it, giving it everything it needs, so why would it need to be hungry?

Ghrelin is the so-called hunger hormone, and—unlike peptide YY and cholecystokinin—it increases appetite. So, if you want to lose weight on a long-term basis, you need to tune down ghrelin. How do you do that? In one study subjects undertook a thirty-three-hour fast, and ghrelin was measured every twenty minutes.

Ghrelin levels are lowest at approximately 9:00 in the morning, the same time that studies of circadian rhythm indicate hunger is lowest. This is also generally the end of the longest period of the day during which you have not had food. This reinforces the fact that hunger is not simply a function of "not having eaten in a while." At 9:00, you have not had food for about fourteen hours, yet you are the least hungry. The surge of counter-regulatory hormones that happens before we wake up also serves to deaden the appetite.

Therefore, again, if hunger is not simply a function of an empty stomach, and is instead a product of our hormones, then eating does not necessarily make you less hungry.

There are three distinct ghrelin peaks, corresponding to lunch, dinner, and the next day's breakfast, suggesting that hunger can be a learned response. We are used to eating three meals per day, so we begin to get hungry just because it is "time to eat." But if you don't eat at those times, ghrelin *does not continually increase.* After the initial wave of hunger, it recedes, and it spontaneously decreases after approximately two hours without food. So, studies suggest that if you ignore your hunger, it will disappear.

If you stop and think about this, you've experienced this diminished ghrelin response before. Think of a time that you were so busy, you worked right through lunch. Maybe at about 1:00 you were hungry, but you just drank some tea, put your nose to the grindstone, and by 3:00 p.m., you were no longer hungry. In those two hours, you "ate" a meal of your own body fat. Your body relied on your stores of food energy to get you through that missed meal. And that's completely natural. That's precisely why we have body fat. You rode a wave of hunger, and it passed because your body took care of itself.

Your average ghrelin levels over twenty-four hours of fasting decrease, meaning that eating nothing over a long period of time makes you *less* hungry. This holds true on very extended fasts, too. A recent study showed that, after three days of a fast, ghrelin and hunger gradually decreased. Yes, you read that right. Test subjects were far *less* hungry when they didn't eat for three days. This jibes perfectly with our clinical experience with clients who are on extended fasts.

Finally, it's worthwhile to note that there's a substantial difference between men and women as far as ghrelin is concerned. When men fast, the hunger hormone decreases only a small amount, but women show a huge decrease in ghrelin. Therefore, you'd expect women to benefit much more from fasting because their hunger drops more. I've found this to be true; many women have told me how a longer fast seemed to completely turn off their cravings.

In conclusion, most people who start fasting are shocked that their hunger was not just manageable, but actually decreased. They usually say something like, "I think my stomach shrank," or "I just can't eat that much anymore." That's perfect. Because if you are no longer as hungry, you are now working with your body to lose weight instead of constantly fighting your body. Intermittent and extended fasting, unlike caloric restriction diets, helps to fix the main problem of weight gain: hunger. Ghrelin, the main hormonal cause of hunger, decreases with fasting, making hunger a manageable problem. In fact, it may not be an issue at all. So, get ready and never fear; you *will* learn to beat the Hunger Bully just as soon as you start fasting. And it will be much, much easier than you think.

MEGAN RAMOS

Have you ever eaten a few pieces of garlic bread, a bowl of pasta, and a bowl of ice cream and still felt hungry? Have you come home from dinner and then wolfed down a bag of popcorn in secret to fill yourself up before bed? You're not alone. I hear these stories from people every day, and I have some of my own. Your mind tells you that you're full because you have to undo the top notch on your belt, but your stomach is still complaining it's empty. These people feel helpless and out of control, bingeing on foods they know won't fill them up.

Then there are the people who are the complete opposite. These are the individuals who eat half a sandwich or a small salad at lunchtime and then declare themselves completely stuffed. And they're not trying to be modest! They *are* full, and they won't eat more, because it is uncomfortable for them to do so.

Many of my clients have undergone bariatric surgery. Their appetites were so far out of control that they felt they needed to be cut open to regulate their bodies. Despite all the promises doctors make that this surgery will help patients lose weight and improve their health, it almost always fails. Initially, most people lose some weight, but after several months the pounds creep back on. Worse, they feel that their appetite is just as out of control as it has ever been. "How can this be?" they ask despairingly. "I've had my stomach physically stapled to make it smaller!"

This is an example of how much we misunderstand the idea of hunger. You don't feel famished because your stomach is so big you can't fill it up. Hunger is also not about your self-control. You can't *will* yourself not to be hungry. You cannot *decide* to be less hungry. You are hungry or you are not. Your appetite is hormonally driven, so hormones are what we need to fix. Not surgically rewiring our intestines. Not counting calories. If you don't regulate your appetite on a hormonal level, you'll never regain control, no matter how small your stomach is. Weight loss, at its core, is not about controlling calories, it's about controlling hunger.

Hunger Is a Habit

I knew I had a problem with hunger when I nearly attacked a woman on a plane for not eating the entire bag of mini pretzels the flight attendant had just given her. I'd devoured my own little bag in less than sixty seconds, and I couldn't figure out how this lady could eat two pretzels and leave the rest sitting there. Confusion, anger, frustration—and most of all, hunger—flooded my body for the rest of the flight, and when the plane landed and I deboarded, I started to cry. I felt pathetic. But there was more to it than that. The clinical, rational part of my brain had gone into overdrive.

What was going on? I was a successful medical researcher, yet I was flipping out over a bag of free pretzels. I'd been disciplined in every other aspect of my life, and there was no reason why I couldn't also be that way about food. Something had to be seriously wrong, and it wasn't lack of willpower or discipline. It wasn't a character flaw. My hunger was a conditioned response. Simply put, it was a bad habit.

If we consistently eat breakfast every single morning at 7:00, lunch at 12:00, and dinner at 6:00 p.m., then we learn to be hungry at those times. Even if we ate a huge meal at lunch, and would not otherwise be hungry at dinnertime, we may still become "hungry" because it is 6:00 p.m. and "time to eat dinner." Young children, who have not yet developed these habits, often refuse food at mealtimes, whereas older children learn these habits and will eat despite not being naturally hungry.

These days, we aren't conditioned to eat just three times per day. Most of us are now snacking or having a meal six or more times per day. At a recent conference I went to, for example, attendees were fed a full breakfast. At 10:30, a midmorning snack was provided, and most of the physician audience nibbled on something. In offices around North America, somebody will bring muffins or bagels to the midmorning or midafternoon meeting. Let's think about this. We just ate. Why do we need to eat again? There is no reason at all. We are building a habit of continually eating, despite the fact that we cannot possibly all be hungry.

Finally, hunger is also a highly suggestible state. That is, we may not be hungry one second, but when we smell hot, delicious, cheesy pizza while walking through the mall food court, we may become quite ravenous. That is a natural stimulus. For me, hearing that little bag of mini pretzels opening was like ringing a dinner bell. I wasn't hungry, but once I thought about food, I couldn't stop dwelling on it. It was a reflex and had nothing to do with discipline or strength of character.

So, the question is: How to combat this? Fasting offers a unique solution. Randomly skipping meals and varying the eating intervals helps break our current habit of feeding three to six times a day. Rather than being hungry just because it is time to eat, we become hungry only when we are really and truly famished. Similarly, by not eating throughout the entire day, we can break ingrained associations between food and a stimulus—watching TV, going to the movies, taking a long car ride, going to your child's sports practice, and so on. For me, being on an airplane was a stimulus—the idea of those little bags of pretzels roused my hunger. By the time the flight attendant came around to serve me, I was already drooling.

Fasting can break all these conditioned responses. If you are not accustomed to eating every two hours, then you will not start salivating like Pavlov's dog every two hours. If we form habits to eat every two hours, then it is no wonder we find it increasingly difficult to resist all the fast-food restaurants while walking around. We are bombarded daily with images of food, references to food, and food stores. The combination of the convenience and availability of food and our ingrained Pavlovian response is deadly for our health.

But going cold turkey is not the most successful method of breaking habits. Research and my clinical experience suggest that a more effective strategy is to replace an unhealthy habit with another, less harmful one. For example, suppose you have a habit of munching on chips or popcorn while watching TV. Simply quitting will make you feel that something is missing. Instead, replace that fattening snack with a cup of herbal or green tea. Yes, you will find this unsatisfying at first, but you will feel a lot less like something is "missing." I discovered over time that I really love jasmine green tea, and eventually I used that to fill my need to consume something. This is the same reason that smokers trying to quit usually chew gum. During fasting, instead of completely skipping lunch (or breakfast), drink a cup of coffee. Try replacing dinner with a bowl of homemade bone broth. Switching habits instead of going cold turkey will be easier in the long run.

Social influences also play a large role in eating habits. When we get together with friends, it is often over a meal, coffee, or a cocktail. This is normal, natural, and part of human culture worldwide. Fighting it is clearly not a winning strategy. Completely avoiding social situations and friends is not healthy, either. What to do? Don't try to fight it. As I'll show you in chapter 20, you can fit the fasting into your schedule.

The day I knew I was healed hormonally from hunger had nothing to do with blood test results, body composition analysis, or dress size. My best friend—who was closer to a sister really—nearly died during labor. While she was recovering and her newborn boy was being monitored in the neonatal ICU, I went to see her, distraught. When I entered the hospital, I had an intense desire for jasmine green tea, so I went to the hospital cafeteria.

As I drank my tea, I realized I was surrounded by all my old comfort foods. There were pretzels, potato chips, bagels, and French fries all around me. But all I wanted was my tea. I didn't care for those other foods nor did I care about the people stuffing their faces at the next table. It had been such a long and hard battle, but I won. I had changed a destructive habit (eating junk foods) to an innocuous one (drinking green tea).

Fasting has given me back control over my body. I can't even describe how much that empowers me. Part of me is still sad from time to time that I had to struggle with this in the first place. Part of me is angry at the world for the state of our entire food system. What calms me down is knowing that there is something I can do about it, and that I can teach others how to do this and succeed, too.

A young doctor I know struggled with obesity for many years but eventually lost some weight with a low-carbohydrate diet. He wasn't at his ideal weight, but he felt happy enough that he was experiencing some success. Unfortunately, though, he still struggled to avoid unhealthy foods.

After spending the week with me and seeing all the ways fasting had benefited my clients, he was motivated to try a seven-day fast, which meant—you got it—not eating for a full week. He started his fast without much difficulty, but he was nervous knowing his problem with hunger. "Don't worry," I said. "Just wait." When I followed up with him after his fast was over, he said to me, smiling, "For the first time in my life I turned down food because I just didn't want it. It wasn't that I was abstaining because I was fasting. I just really wasn't hungry. My appetite decreased! Megan, I've never turned down food like that before."

Every day I see clients like this. Clients come into my office and cry because for the first time in years they feel in control of their bodies. I can see the change in how they carry themselves. They stand taller, with their chests more upright. They hold their head higher. Even their eyes seem clearer. Witnessing this kind of change is the best part of my day.

How to Break the Hunger Habit

Hunger is often associated with the time of day you're used to eating, a place, or an occasion. Here are a few easy ways to break a conditioned response to hunger.

1. Eat only at the table. No eating at your desk. No eating in the car. No eating on the couch. No eating in bed. No eating during class. No eating at the movies or at a sporting event.
2. If you're tempted to eat at a certain time of day, like 3:30 p.m. because you always snack at 3:30, set an alarm for that time. When the alarm goes off, drink a glass of water or a cup of tea instead of eating. Chances are, you'll feel full.
3. When you're on a plane, put on headphones when the flight attendant walks by and decline offers of snacks.

CHAPTER 4

Forget Calorie Restriction

EVE MAYER

I can't tell you how many times I've heard someone repeat some version of, "Put the fork down, honey, and you'll lose that weight!"

If you're like me, you've tried to eat less a thousand times, and each and every time you've failed. I've dabbled in the art of dieting since I was eight, when I first started worrying about getting chunky, and pretty much all these efforts involved some version of cutting calories.

Other Ways to Say "Calorie Restriction"
- Calories in versus calories out.
- Eating less or eating smaller meals more often.
- The energy balance equation.
- Portion control.
- In scientific terms, the Law of Thermodynamics.

My first real diet was probably the most ridiculous version of calorie restriction you can imagine. When I was in my early teens, I'd put on an extra fifteen pounds. After carefully considering what food always filled me up, I proudly designed a diet around it. Ladies and gentlemen, my creation was called "the Candy Bar Diet."

I skipped breakfast and lunch, and at the dinner table every night, I tricked my mom by pushing food around on my plate and eating a bite or two when I wasn't talking. At school in the afternoon, when my stomach had really started rumbling, I'd creep into the stairwell where the vending machine stood, and I'd secretly purchase one Kit Kat. I'd peel back the paper from the Kit Kat and make my delicious "meal" last twenty minutes by biting off the outside chocolate and pulling apart each of the individual layers one by one to enjoy them separately. I knew that if I could eat 1,000 calories a day, I would lose weight. A Kit Kat only contained 218 calories, so I would be skinny in no time!

The Kit Kat diet lasted one week, and I lost six pounds. After one week of feeling like crud and being hungry all the time, I officially gave it up, and I went back to eating as I had before. I gained nine pounds in the following two weeks.

The idea of going on a diet always started the same way for me. I'd look down at my rolls of belly fat or feel my thighs rubbing against each other, and I'd think about the promise of a new life. I'd plan for a day or two, then hit the ground running. Over the years, I tried SlimFast, Nutrisystem, Weight Watchers, the Cabbage Soup Diet, and any number of juice cleanses. I found diets in magazines, featuring glossy photos of portion-controlled plates, I picked up diets designed by spas and health organizations, and I followed a diet program created by my doctor. I've been on at least fifty diets that have restricted how many calories I eat, and I've counted fat, sodium, water, and macros, and almost any other nutrient there is. I've eaten 200, 600, 800, 1,200, or 1,800 calories a day, and I've always lost weight at first. Yet I've always felt grumpy, tired, angry, frustrated, distracted, and miserable doing it.

The less I consumed, the hungrier I felt, and the more times I failed to lose weight, the more I reinforced the belief that I was a failure. Feeling hopeless, I'd eat whatever I wanted when I was off my diet, and not only did I gain the weight back, but I added a few extra pounds for good measure. I call this my "Big, Fat Circle of Despair" because it's a dangerous ride I can never safely escape from, and the only time I felt happy is in the day or two before I started a new diet.

I didn't know it a few years ago, before I started fasting, but calorie restriction is a lie. Through all the miserable days of my many failed diets, I was constantly hungry and unable to break free from obesity, yet I had no idea why.

MEGAN RAMOS

Why? Because the notion of calories is overly simplistic, and the theory behind calorie restriction is completely wrong. Let me break it down for you.

Calories and Metabolism

The famous "Calories In, Calories Out" (CICO) hypothesis of obesity—more formally called the Energy Balance Equation—comes down to a simple formula: Fat Accumulation = Calories

In—Calories Out. We spend most of our time obsessing over "Calories In" without giving much consideration to "Calories Out," other than exercise. The reason most people don't think of calorie expenditure in more complex terms is that it's tough to wrap your brain around your body's basal metabolic rate, or BMR. As we discussed previously, BMR is the amount of energy (calories) your body burns just to stay alive. BMR does not include exercise; it's solely the energy required by your organs to function. BMR is hard to measure, so most "experts" assume it doesn't change.

This is completely untrue. BMR can go up or down by 50 percent depending on your caloric intake, as well as other factors like breathing and total energy expenditure.

That's why calorie-restriction diets stop working. When you cut calories, your basal metabolic rate begins to slow down. You've reduced your calories in, but in response, your body has reduced "Calories Out." Your organ systems start making budgetary energy cuts across the board—a little bit from your reproductive system, a little bit from your respiratory system, a little bit from your cognitive function—and soon you're expending fewer calories. For example, if you eat 500 fewer calories, but your body burns 500 fewer calories, you will not lose body fat. This is the dirty little secret of calorie-restriction diets.

I don't know a single person who hasn't tried one or more of these diets, and most, if not all, have lost weight initially, then plateaued. Like Eve, they feel hungrier with each passing day, and, because their bodies are burning fewer calories, they get cold and tired, too. When their metabolic rate hits rock bottom, these people redouble their efforts, which works, but only for a short period of time. Then they give up, convinced their body is broken, and go back to their old way of eating. Within a few weeks they weigh more than before they started. A few months or a year after that, they go on a different diet, but it's another version of caloric restriction, and it entirely ignores the idea of BMR.

Sound familiar?

Let me show you a better way to think about calories.

Calories Differ

When I meet with a new client, I start by asking them: *What are five non-junk foods you should cut out of your diet to lose weight?* They generally respond by naming bread, rice, pasta, potatoes, and corn. I then explain that, while these foods are definitely fattening, they're also extremely low in fat and calories. You read that right: the foods that can make you gain weight are *also* low in calories.

Most of my clients initially seem shocked, but I don't explain anything to them right away. Instead, I let it sink in that much of the nutritional information they've heard for years is completely wrong. Then I ask them two further questions:

Which is more likely to make you gain weight: a can of soda or a handful of raw almonds?

Without fail, my clients suddenly perk up because they feel they can confidently answer. Of course, they choose "soda." Then, I ask:

Is a handful of raw almonds good or bad for weight loss?

Almost every person says that the answer is "good."

This is when I go in for the kill!

A can of soda is 160 calories, and it's likely to cause weight gain. But a handful of raw almonds is *also* 160 calories, and it's likely to lead to weight loss. How can the same number of calories do two totally different things? If calories differ so much, how do they affect weight gain or loss? And, if it doesn't really matter what you eat, does it follow that a person can live on soda—or, if you're Eve, Kit Kats—and *still* lose weight?

Of course not. The way our bodies process food is more complex than that.

Our body's hormonal response to two food items containing equal calories is measurably different depending on the composition of the food. The sugar in a soda will spike your blood sugar levels and cause your pancreas to produce huge amounts of insulin. The nutrients in the almonds—which include protein and fat—will not. Once almonds are digested, your body experiences only a small increase in its blood glucose levels.

Therefore, it goes to figure that a calorie is a calorie, just like a dog is a dog, but there are many different breeds of dogs. Your body's hormonal and metabolic responses to the soda and almonds are completely different, and therefore the fattening effects of those foods differ as well. In addition, every single bodily function—from blood sugar control to body temperature—is regulated by a network of hormonal systems. The "Calories In, Calories Out" hypothesis of obesity asks us to assume that fat cells are unregulated. Why would our hormones control everything in our body but fat cells? This thinking doesn't make any sense.

So, if calories don't matter, how do we gain weight? The answer to this question shouldn't be a surprise to any doctor who has ever prescribed insulin to a patient with type 2 diabetes. As soon as the person starts taking insulin, they gain weight, and they put on more and more each time their dose increases. It doesn't matter if they eat less or move more; they'll gain weight regardless.

Therefore, the answer to weight regulation lies in insulin.

How Fasting Is Different

The common dietary advice to avoid fat and eat small meals to boost your metabolism did nothing to control your hunger. For years you've been told that you lack the willpower to eat less, and that being overweight is your fault. Nothing could be farther from the truth. It's *calorie-restriction diets* that cause your metabolism to slow down to a crawl.

Fasting does just the opposite. The hormone noradrenaline, which is produced during a period of fasting, boosts metabolic rates. As you fast more and regularly over time, metabolic rates go up, and more weight loss is achieved.

For example, we worked with a forty-seven-year-old woman who'd tried every popular diet in the United States, yet ended up having a BMR of 487 cal/day.* This is *extremely* low. A 135-pound, 5-foot, 5-inch forty-seven-year-old woman should have a BMR of around 1,200. After six months of intermittent fasting and following a low-carb, high-fat diet, this client was able to raise her BMR to about 800. A year later, it went up to around 1,200. She didn't start to lose weight until six months of following this new lifestyle, but she understood the science and knew fasting and her new diet were her solution. She was patient, and I'm so glad she was! Close to the eight-to-twelve-month mark she really started to lose weight, and, as of this date, she's down sixty pounds. She *never* could have done that on a calorie-reduction diet!

---

*   BMR measurements are often taken by a health professional. To complete them, measurements of carbon dioxide and oxygen levels are recorded after a person fasts for twelve hours and gets at least eight hours of sleep. Your ideal BMR can be measured roughly using this formula:

For men: BMR = 10 × weight (kg) + 6.25 × height (cm) – 5 × age (years) + 5

For women: BMR = 10 × weight (kg) + 6.25 × height (cm) – 5 × age (years) – 161

JASON FUNG

As I argued in my book *The Obesity Code,* the obsessive fixation on calories is incorrect. Let's think about this. Up until the 1970s, there was little obesity, yet people had virtually no idea how many calories they ate. They also didn't pay attention to how many calories they burned. Exercise was not something you did for fun, yet people all around the world were far healthier and slimmer than they are today. If the majority of the world's people were able to avoid obesity without counting calories, then how is counting calories so fundamental to weight stability today? Our society wasn't grossly overweight for over five thousand years, but since 1980, we can't live without calorie and step counters? It doesn't make sense.

There have been two main changes to the American diet since the 1970s. First, we were advised to lower the fat in our diets and increase the amount of carbohydrates we consume. This directive to eat more bread and pasta turned out *not* to be particularly slimming. But there's also another problem that has largely flown under the radar: the increase in meal frequency.

Our Culture of Overconsumption

Before the 1980s, people typically ate three times per day: breakfast, lunch, and dinner. If you were not hungry, then it was perfectly acceptable to skip a meal. By 2004, the number of meals per day had increased closer to six—almost double. Now, snacking was not just an indulgence, it was encouraged as a healthy behavior. Meal skipping was heavily frowned upon. What kind of strange world was this, where you need to eat constantly to lose weight?

Doctors, dieticians, and magazines told people to never, ever skip a meal, warning of dire consequences. But what happens when you don't eat that is so bad? Let's see. If you don't eat, your body burns stored body fat in order to get the energy it needs. That's all. We store fat so that we can use it, and if we don't eat, we use the body fat.

Retail and culture—societal, familial, and business—have mushroomed around this habit of overconsumption. Today, there are dozens of eating opportunities in a single day, including breakfast, midmorning meetings where somebody orders muffins and bagels, lunch, midafternoon coffee breaks, dinner, after-dinner drinks (with snacks), and then time to curl up in front of the TV with some more snacks.

This is a relatively new phenomenon and is the product of a culture where cheap, low-quality food products are mass produced. If you think about it, doesn't it seem strange that consuming is the norm, and abstaining from food is something that takes effort? Consider what it would take to eat a doughnut in the past. If you lived on a farm in nineteenth-century America, you would need to plant some wheat, raise some cows to produce milk, and buy some sugar. Six months later, you could harvest the wheat, grind it into flour, mix it together with your milk and sugar, then spend a few more hours cooking it in order to enjoy your tasty treat. With all the work involved, you probably wouldn't bother, right?

Not these days. There's a Dunkin' Donuts at every highway exit. There's even a doughnut shop in the hospital where I work.

What about the 1970s, before the obesity epidemic? It was much easier to get a doughnut than it was in the 1800s, but it was still not socially acceptable. If you asked for an after-school snack, your mom would usually say, "No, you'll ruin your dinner." If you wanted a bedtime snack, she said, "No, you should have eaten more at dinner." If you brought a muffin to your midafternoon meeting at work, your colleagues would look at you like you were rude or crazy. If you were a naughty child and sent to bed without dinner, nobody worried that missing a single meal would irrevocably ruin your health. Dieticians, doctors, and all other nutritional professionals were quite clear that we should eat three square meals a day.

All that's changed.

Furthermore, the body of people who used to understand and support fasting has moved to the fringe. In the past, if you were to fast, you would do it within a community of like-minded people. For example, if you were Catholic, during Lent, your priest would discuss fasting, and your family and friends at church would fast with you. There wouldn't be plates of food on every conference-room table or snack machines in every hall, and you would not be expected to cook because nobody was eating. The same applies to Muslims, Jews, Buddhists, Hindus, and others during their particular periods of fasting.

Under these conditions, fasting was not so difficult. Sure, you might have been hungry, but knowing that everybody else was feeling the same was comforting. There were no foods to tempt you to "cheat," and while you'd understand that fasting was uncomfortable, that was the point of Lent. It was something you did naturally at that time of the year. Sometimes you feast, and sometimes you fast.

Fast-forward to the present day. Snacking is actively encouraged as healthy, even as we groan under the weight of an obesity crisis. If we don't give our children snacks, people practically consider that child abuse. It is socially acceptable to eat in your car, at your desk, while walking, while talking on the phone, in the movie theater. We've put cup holders into our cars to make eating there easier. Doctors, nutritionists, and dieticians will tell you to make sure you eat constantly throughout the day or else you'll go into "starvation mode," and they argue that skipping meals will make you fat.

We often think that we're in control of the decisions we make, but behavioral psychology suggests that motivation comes, in larger part, from societal influences. The Nobel Prize–winning work of Daniel Kahneman and Richard Thaler in behavioral economics showed that humans are, as Dan Ariely puts it, predictably irrational.

Consider the example of organ donation rates.

In Denmark there is about a 4 percent rate of organ donation. In neighboring Sweden, that rate is 89 percent. The Danes and Swedes are very much similar in almost all respects, so why the huge discrepancy? The answer lies in the default state. In Denmark, you check the box if you would like to opt *into* the organ donor program. In Sweden, you check the box if you would like to opt *out* of the organ donor program. The difference does not lie in the people or their values.

Take this example, too. Recently I signed up for a free trial of Amazon Prime and was automatically enrolled. Long after it ceased to be beneficial for me, I'm still a member. This phenomenon, of course, is well known. If a problem is too complex or overwhelming, then inertia takes over. When we don't know what to do, we simply take the choice that's already been made for us.

So, how is weight loss automatic in the 1970s and weight gain automatic in the 2000s? The problem is not the people; the problem is the system or default. The biggest issue is that we see obesity as a people problem instead of a system problem. For example, consider the logic that would suggest that if more people are obese today than in the 1970s, then people today have less willpower. Does that even make any sense? This is what leads to fat-shaming.

The major difference between the 1970s and today is that the default today is "eating," while the default in the 1970s was "not eating." This, just like the organ donation issue, has overwhelming implications. Luckily, it doesn't take any willpower to change this model of behavior. It is a matter of fixing the default. You need to change your environment, which is a lot easier than changing yourself. If you set the default to "not eating," then it takes willpower to eat.

Let's look at an example of a midafternoon meeting. You're bored and not particularly hungry, but you start to think about a delicious chocolate chip cookie. Do you suddenly and rudely leave the meeting, get into your car, drive to the local bakery, and buy a cookie? Do you walk back into the meeting, with crumbs on your shirt, as your colleagues stare in silent, horrified disapproval? Of course you wouldn't do this. But what if the office manager had ordered a plate of cookies and coffee for the meeting? Perhaps 90 percent of us would mindlessly eat that cookie placed temptingly in the room. The difference between you eating and not eating? Just as with the Danish and Swedish donation rates, it is the default state. You can go get a snack anytime you want, but it's mandatory to have it sitting in front of you during meetings. If you fix the default, then weight loss will become automatic. Intermittent fasting, of course, sets the default to "not eating" and helps people lose weight. "Not eating" becomes the new "eating."

I believe there's a large degree of comfort in changing our perspective about the amount and kinds of foods we eat—and when and how often we consume them. If we look at our environment differently, recognizing that eating all the time is *not* normal, it takes a lot of the pressure and choice out of high-stress food environments. The burden of constantly needing to eat is gone. The anxiety about continuously shopping and stopping to snack is absent. Hunger is no longer an imposition; it's a natural part of our body chemistry. At first, fasting may feel challenging, but it ultimately liberates your body, your mind, and your lifestyle.

CHAPTER 5

A Path to Healthier Eating

EVE MAYER

One of the surest ways to succeed with fasting is to have a clear picture of what you're planning to consume. That's right. Weight loss while you're fasting is as much about what you *are* eating as what you aren't.

Interestingly, this idea would seem to fly in the face of some research. There are now numerous medical studies that show that fasting works even when a person doesn't change their diet. This research says that fasting *on its own* can help to lower blood sugar levels and assist in weight loss—regardless of what you eat. I've met a few people who are proof of this, including Sally, a woman in her late twenties who uses intermittent fasting to keep her weight steady. She also does cardio exercise and strength training three times a week. Sally is in great shape and has always consumed anything she wanted, including doughnuts, cake, sweet potatoes, steak, and fried chicken. However, Sally explained to me that she eats what she likes but does so very strategically, often only eating once or twice a day and never after 8:00 p.m.

I also spoke with a sixty-four-year-old gentleman who has repeatedly tried to cut down his sugar intake—especially ice cream and cookies—and found it impossible. However, Jack is able to say no to most sweets when he does intermittent fasting, finding it easier to abstain *entirely* from snacks than to pass on sweets when he *is* eating. Jack began skipping breakfast and eating twice daily, at noon and 7:00 p.m., and, in two months, he dropped twenty-five pounds and went off his blood pressure medicine.

So, if it's been proved that fasting without changing what you eat can lead to success, then why are we going to talk about food? Because most of the people I meet aren't like Jack and Sally. Most of the individuals I've come across have struggled with poor food choices most of their lives, and while fasting alone can provide *some* success for them, eating better can lead to better health outcomes. Indeed, combining a healthier diet with fasting is likely to give most people *greater success that lasts longer.*

This has been the case for me. Gone are my days of endless gumbo and beignets. I stay away from carbs 90 percent of the time these days and eat mostly protein or fats in the form of avocados, cheese, meat, and seafood, plus green vegetables and occasionally berries. The other 10 percent of the time I allow myself to eat whatever I want, including cake, breaded or fried foods, chocolate, fruit, and grains. When I stick to a high-and-healthy-fat and low-carb way of eating 90 percent of the time, I can count on my weight to either decrease or stay steady. My body also feels strong, and I rarely get headaches or sinus infections, which I used to experience frequently when my diet was poor. I tend to feel hungry four or five times a day, which is wonderful because, for most of my life, I felt starving almost every minute of every day. Upping my intake of high-fat foods has also helped me to stay fuller longer.

Now is the time for *you* to decide what foods fill you up and create the best chance of health in your body. Megan will offer you clear, easy-to-follow advice about how low-carb/healthy-fat options can make a fasting lifestyle easier for you. But you will have to choose and test these foods for yourself, paying attention to what is best for you.

MEGAN RAMOS

The science is clear: the inability to lose weight, and all the health complaints that may surround it—including cardiovascular disease, type 2 diabetes, certain cancers, stroke, metabolic syndrome, PCOS, and more—signal a hormonal instability. The foods you choose to eat may raise your blood sugar and insulin levels, inhibiting fat-burning and suppressing your body's ability to signal that it's full. Fasting *is* going to help with these problems, but, like Eve said, what you put on your plate is often just as important as what and when you don't.

While we won't offer a set eating plan or step-by-step diet in this chapter, Jason, Eve, and I—as well as many doctors and researchers—are strongly of the opinion that sticking to a low-carb diet that's high in healthy fats will help keep you in the fasting lane. How often you need to follow it is up to you, and everyone is different. Eve maintains this diet 90 percent of the time. Other clients of mine can't *go near* an unprocessed carb—

ever. Here, we'll simply paint the broad brushstrokes, giving you an overview of the benefits of a low-carb, healthy-fat lifestyle and offering a few categories of foods from which to choose.

What Is Low-Carb?

Carbohydrates are compounds consisting of sugar, starch, and fiber. Found in all kinds of foods—from potatoes to bread to rice to soda—they can be an abundant source of energy for the human body. Contrary to the food pyramid, however, they are not a *necessary* source. In the absence of them, our bodies will convert protein into glucose for the very few of its parts that may need it.

There are two types of carbs: refined and unrefined carbs. Refined carbs contain two sugar molecules, and they have been stripped of bran, fiber, and nutrients. These include pasta and pure sugar, and they cause a blood sugar spike. When they hit your stomach, the pancreas increases insulin secretion so that they can be converted quickly to glucose.

Unrefined carbohydrates—found in whole grains, beans, potatoes, and more—are made up of longer sugar chains. The body burns them more slowly than it does refined carbs, but they still cause the pancreas to secrete a surge of insulin. They still cause blood sugar to rise.

As we've previously discussed, persistent high glucose levels can lead to type 2 diabetes, as well as increase your risk of heart disease, stroke, and more. Since your body quickly burns the glucose that carbs help produce, it has less of a chance to enter fat-burning mode. With carbs, a sudden rise in blood sugar leads to an equally sudden drop, so you're hungry. You're tempted to eat more. You do eat more, and more frequently. You get fat.

But what exactly is low-carb? What amount should a person aim for? While there's no set definition, and every diet plan will tell you something different, a liberal low-carb diet is generally considered to consist of 50 to 100 grams of carbohydrates a day. A moderate low-carb diet is between 21 and 50 grams a day, while a strict, ketogenic diet is 20 grams or less a day. In the Western world, men typically eat about 200 to 330 grams of carbs a day, while women eat around 180 to 230 grams daily. These numbers truly illustrate how profound our culture of excess carb consumption is, and they demonstrate the fact that most people need to cut down their carbs.

We don't recommend counting carbs obsessively, though. Not only is it time-consuming, it also causes you to focus on numbers rather than the quality of the food you're eating. A low-carb diet is not one that includes three Milky Way bars a day (40 grams of carbs each), with no other carbohydrate source. Simply avoid the foods on the following list, focusing instead on dairy foods, fish and seafood, poultry, red meat, vegetables that grow above ground, and nuts. Together, these will give you a healthy serving of carbs without causing you to go overboard.

Avoid These Carbs!
Candy: 70 grams of carbs*
Doughnuts: 49 grams

In general, avoid *all* products containing refined (white) sugar. This includes sports drinks, soda, cakes, cookies, ice cream, breakfast cereal, muffins, and more.

White bread: 46 grams
Cooked pasta: 29 grams

Avoid all starches. These include bread, buns, pasta, and anything with flour—including whole-wheat flour. It's often tough to determine which starchy foods contain unrefined carbs versus refined carbs, so take out the worry and avoid them all. In addition, "gluten-free" does not mean "carb-free," so steer clear of these products.

Cooked rice: 28 grams

That's right. Rice is high-carb, even brown rice, which hasn't been bleached, processed, and stripped of its nutrients. Avoid rice altogether on a low-carb diet.

Potatoes: 15 grams

This includes all varieties of potatoes, potato chips, and French fries.

Beans: It's a surprise to many people, but beans are high in carbs, so it's best to avoid them on a low-carb diet unless you are a vegetarian or vegan.

Fruit: Berries like blueberries, raspberries, and strawberries are typically fine to eat about once a day. Avoid other fruits. Bananas, mangoes, and oranges may be full of vitamins, but they're high in carbs, which raise your blood sugar.

* All these measurements are per 3.5-gram serving.

The Glycemic Index and Glycemic Load

To fully understand the impact of carbs on your blood sugar, you'll need to understand two dietary terms: glycemic index (GI) and glycemic load (GL). GI is a measure of how quickly 50 grams of a food containing carbohydrates is digested and begins to impact your blood sugar levels. Every food is scored, 1 to 100, on the glycemic index, with the lower number indicating foods that don't cause a spike in blood sugar and the higher numbers being those that do. (To determine the GI index of a food, simply search online. The value is rarely on a food label.) Not surprisingly, many carb-rich foods score high on the glycemic index.

What matters in terms of blood sugar levels, though, is not just the GI value of the food, but also how much of it you eat. For some foods, 50 grams is far more than a typical serving, and for others, it is less. Taken together, the GI value and the portion are called the glycemic load (GL), which indicates how high your insulin levels will rise and how long they'll stay that way. The way to calculate that number is:

GL = (GI × grams of carbohydrate) divided by 100.

For example, an apple has a GI of 38 and contains 13 grams of carbohydrates, so its GL = (38 x 13)/100 = 5. A Snickers has a GI of 55 and contains 64 grams of carbs, so its GL = (55 x 64)/100 = 35. I'm sure you didn't need the GL value to tell you this, but a Snickers is going to spike your blood sugar, then cause it to crash a lot faster than an apple will.

Staying on a low-GL diet is not just about losing weight, either. Carbs like oatmeal and barley—which have a GL value under 20—will help reduce your risk of diabetes, heart disease, and certain cancers.

Great Low-GL Foods to Enjoy While Fasting

These foods contain a glycemic load of 10 or less and are great for incorporating into any of your meals. Or, you can eat them on their own!

- Carrots
- Nuts
- Meats and seafood
- Berries
- Plain yogurt
- Cheese

The Importance of Fats

When I was growing up, dietary fat was demonized. Most people believed that foods containing high amounts of fat made you gain weight, so butter, cheese, oil, eggs, red meat, and anything greasy was off-limits if you cared anything about your health. Fats became enemy #1, due, in part, to a misunderstanding about the role fat plays in cardiovascular health. People believed that—across the board—fat raises your cholesterol levels. Cholesterol clogs your arteries. Clogged arteries lead to heart attacks. Heart attacks kill.

This has been the prevailing wisdom since the 1960s, and it is just plain wrong.

There are two types of fat: saturated and unsaturated. Saturated fats, which are fats that are solid at room temperature, can be found in animal meat, cheeses, and butter. Unsaturated fats are liquid at room temperature, and they include oils like avocado, canola, and olive oil. Saturated fats elevate low-density lipoprotein (LDL) cholesterol levels, so people used to fear that this would lead to clogged arteries, then to heart disease. But a 2005 study that followed 235 women with established heart disease over three years found that there was no relationship between fat intake and the increase of arterial blockages. In fact, the women eating *the most saturated fat* saw a regression in their blockages over the women who ate less fat!

This study wasn't alone in showing that saturated fats aren't bad for you. A 2010 Japanese study followed 58,543 men and women over 14.1 years and discovered that eating more saturated fat helped protect test subjects against heart attack and stroke. We've been getting the wrong information for years! All kinds of fats—including saturated fats—are actually *good* for you.

Where can you find the best sources of fats? I recommend that you choose a diverse range of foods from the following categories:

- **MEAT:** Choose beef, pork, lamb, game, or poultry. You can eat the fat on the meat and skin on the chicken. Try to choose grass-fed meat, as it contains fewer hormones.
- **FISH AND SEAFOOD:** Any kind, but especially salmon, mackerel, sardines, and herring because they contain so many omega-3 fatty acids.
- **EGGS:** Cook them however you like.
- **FATS FOR COOKING:** Coconut, olive oil, butter, ghee (clarified butter used in Southeast Asian, Middle Eastern, and Indian cuisine), avocado oil, and beef tallow are all healthy options.
- **DAIRY:** Choose full-fat dairy, yogurt, and unprocessed cheeses.
- **NUTS:** Macadamia, pine, Brazil, walnuts, and almonds. Avoid peanuts and cashews.
- **FRUITS:** Olives and avocados. Eat berries no more than once a day.

Ketosis and Fasting

Ketosis is a state when the body stops burning glucose and starts burning fat. It is a perfectly normal metabolic process that happens during fasting after approximately twenty-four hours without food. Ketones cross the blood-brain barrier and give the brain much of the energy it needs to function. Ketones are never used by any other organ in the body, or by the muscles, but to the brain, they're essential during fasting.

There has been some confusion between ketosis and fat adaption, so allow me to clear that up. Fat adaption is a process that happens after at least four weeks of ketosis. It's the end of your period of transition into a low-carb diet. During fat adaption, your body is fully acclimated to burning only fat. You no longer crave carbs, and you tend to get fuller faster and stay that way longer. When you eat carbs, they don't spike your blood sugar the way they used to, and blood sugar returns to a normal state more quickly. Your body may even become fat adapted without your even knowing it.

Other Diet Considerations

A good diet while fasting is all about balance. A person shouldn't intentionally add a fat to a meal, thinking it won't matter. For example, don't put extra butter on your pork chop just because you know that eating fat can be healthy. Yes, the fat in the pork is good, as is the fat in the butter, but your body doesn't *need* the extra. Your body won't burn extra fat simply because it exists.

Vegetables should also be part of your meals. Add them liberally and try a variety. Vegetables that grow above the ground, including cauliflower, broccoli, cabbage, Brussels sprouts, kale, collards, bok choy, spinach, asparagus, zucchini, eggplant, mushrooms, cucumber, onions, peppers, and leafy greens (including lettuce) are full of vitamins and minerals your body needs. They're also super-low in carbs. Eat avocados and olives liberally—they're loaded with healthy fats—but consume berries no more than once a day, when you have a craving for fruit.

Stay away from processed sugar and processed foods. I can't say this enough. No candy, cookies, chips, or sodas. If you look at packaging and see that something has dozens of ingredients in it, don't eat it. Choose fresh, whole ingredients whenever possible, and drink plenty of water whenever you're thirsty. We often mistake thirst for hunger, so keep water on hand at all times; staying hydrated will help keep hunger pangs away.

CHAPTER 6

Prepare to Think Differently About Food

EVE MAYER

Globs of warm, crystallized sugar spread over my tongue like a blanket as I crunched the pecan bits. My eyes rolled back in my head with a pleasure so pure it could rival finding a pile of Christmas presents under the tree.

I was seven years old, and I was licking the spoon from the pot from my mom's pralines.

A praline is one of Louisiana's most famous sweets, made of sugar, evaporated milk, butter, pecans, and a few secret ingredients. You stir this sweet nectar of the Cajun gods in a heavy pot until it gets to the precise temperature and emits a fragrance beyond your wildest candy land dreams. My mom's pralines were so good that she started a candy company, and many nights after my parents got home from work and I'd completed my homework, we'd sit around the table together bagging and labeling pralines for sale and packaging them to be shipped.

These nights together—with pralines crumbling on the table and me eating the broken bits—meant family. Food wasn't a way for my body to generate energy, it was the height of pleasure and time together with the people I loved most. For me, food was emotionally charged—full of excitement in the good times and bitter comfort when my mom got sick—and, together, those deeply ingrained beliefs made me fat. When I began fasting, I had to let go of them, so I decided to start thinking of food differently.

Food Is for Energy

The first step in reframing how you think about food is to see it as a potential source of energy. Full stop. Food can (and should!) taste good to you, and while there's no reason that my aunt Rose's deviled eggs, Lebanese kibbeh, and roast shouldn't have been a part of every family gathering, it didn't help me to think of food as love. Every delicious praline crumb did *not* make family conversation better, and every morsel in my mouth did not communicate my thanks to the chef. When I started fasting, I consciously and deliberately had to think of food as energy and nothing more, divorcing it from the more complicated places it had always stood for in my mind.

I also had to tell myself—repeatedly—that I already had plenty of energy stored as fat, and that fasting would help me burn it.

• • •

What Does Food Mean to You?

For so many of us, food is a reward. In my family, every time something went well—an A on a test, a good run at softball, a solo in band, even beautiful weather—we stuffed our faces. Food was always there to be my friend, to help me celebrate, and to accentuate my joy when things were good.

It was never a fickle friend, either. It also showed up when things went badly, like when my dad and I sat in the National Institutes of Health in Bethesda, Maryland, and he gave me a half smile, hugged me, and blinked back tears as he told me they were testing my mom to figure out why she was so sick. We stopped for lunch in the cafeteria that day, and I ate the most delicious, juicy cheeseburger, piled high with provolone cheese. That hamburger whisked away an eight-year-old's emotions and protected me from my greatest fears.

For many others just like me, food is there in times of boredom or loneliness, a friend or a diversion to fill the gaps in one's heart or one's schedule. When I started college, I realized I was sick of being in every club, which I'd done in high school, and I was tired of pushing myself to my limits. I just wanted to be lazy, and food was there to be my friend. I ate to keep myself occupied—my hands busy and my mind entertained—alone or with friends. In two semesters, I gained 50 pounds. As a 225-pound twenty-year-old, I found dating tough. My self-confidence plummeted, and I started dating men who weren't right for me because I believed they were my only options. When things inevitably didn't work out, food was there for me, and there was no pain a cupcake couldn't conquer. By the time I was an adult, I'd swelled to a size 26, at 300 pounds.

Food is certainly not the enemy, but it also shouldn't be leaned on as a best friend, a comfort, a reward, a distraction. How can you begin to explore what food means to you? One way is to recall the times you've overindulged in an extreme way. Was it at your best friend's wedding, when you were so happy you ate three pieces of cake? Or was it after your beloved grandma's funeral, when the crab dip at the appetizer table was the only thing standing between you and despair? Or maybe it's much simpler, like when you're sitting at your desk working, and you want something to distract you when your mind needs to take a break from whatever project you're busy tackling. If you're like me, these moments when food has been your emotional support are too many to remember—or, perhaps, they're just too painful. I'll make this task easier for you:

Grab a notebook on a Sunday and bookmark seven pages, labeling them Monday–Sunday. The following day, begin writing down all the foods and beverages you consume in that day, including breakfast, lunch, dinner, and everything in between. Write down *everything*. Next to each meal or snack, note how you felt when you ate or drank and why you ate or drank. Write down every emotion or state of being—and yes, they can include the obvious one: hunger. You should be sure to log your notes right after eating, so you may want to use a small notebook that you can keep in your bag. Or, perhaps, you can have several notebooks in your kitchen, your car, and your office. Your phone works, too! Try to avoid the temptation to go back and look at the previous day or days' pages the next day. Stay honest—no editing!

At the end of the week, compile the emotions you felt onto one list. Edit the content as needed, so "Ate three Oreos when I was mentally tired and unproductive. Felt hyper after," may become "Mentally slow. Listless. Instant but short-term energy." As you begin to implement fasting, you can work on identifying and processing these feelings as they connect to food so that you can take the emotion out of eating.

Find What Feels Good

Choosing the right healthy foods while you're fasting is important, but it's also vital to listen to your body and follow its cues in terms of the foods that make you feel full and energized, versus those that make you sluggish or slow down your digestion. I've found it helps me to keep lists on my phone that I can add to whenever I want. I have one list for foods that make me feel good and another for foods that make me feel poorly. The more you add to it, the sooner you will see patterns, and you can then create shopping lists full of foods that will make you feel great.

After a month of keeping these two lists, I want you to review them. Keep the list of foods that make you feel poorly and rename it, "Is It Worth It?" You can test out these foods again to see if your taste for them—or how you react to them—has changed. For example, rice was on my "Is It Worth It?" list, but recently I was in Japan, and I chose to eat a small bowl because I was very curious to see if it was as good as I expected. I decided that the possibility of not feeling great after I ate it was an acceptable risk. Lucky for me, the rice was delicious, and it didn't make me feel bad! Now I make rice a part of my diet every now and then. On the other hand, I recently ate cotton candy because it used to be one of my very favorite treats. After not having cotton candy for a year (and very little sugar, for that matter), I was shocked to find that I didn't like the taste of it! I can remember a time when I had eaten four bags of cotton candy in one sitting, but now, I don't want even a bite—all thanks to cutting it out of my life.

My List of Foods

Foods that make me feel good:

Lettuce, bacon, cheese, steak, salmon, avocado, tomato, grass-
fed ground beef, eggs, sausage, chicken, broccoli, raspberries,
dill pickles, tuna fish, ribs, cabbage, stevia-sweetened dark
chocolate.

Foods that make me feel bad:

Ice cream, cotton candy, potatoes, bread, bananas, flour, sweet
potatoes, heavy cream, peaches, stevia-sweetened key lime pie.

If you don't have many items on your list that you truly enjoy,
then it is time to try a wider variety of foods. You are not here to
punish yourself. Eating should *always* be an enjoyable activity.

MEGAN RAMOS

Food plays a role in every celebration in every culture in the history of the world. Holidays call for midday feasts. Birthdays are
celebrated with cake. Backyard barbecues, tailgates, neighborhood potlucks . . . food is used to create community and connection. This is totally okay, and the good news is that fasting doesn't
require you to skip these special occasions. But a problem arises
if we are feasting or eating cake every day instead of once in a
while, because that's when we've crossed the line from simple enjoyment of food to addiction.

When Enjoyment Becomes Addiction

There is a very fine line between enjoyment and addiction. While people have enjoyed eating since the dawn of time, food addiction is almost entirely a modern problem.

Over the course of our evolution on this planet, we developed a taste for foods that are nourishing, give us energy, and can sustain us both throughout the day and in the long term. We also learned how to *stop* eating. If our caveman and cavewoman ancestors became too fat, they wouldn't survive. They wouldn't be able to catch prey, and they couldn't run away from predators like lions, tigers, and bears. The survival of our species has always depended upon eating enough and not eating too much—not eating simply because food is available to us.

When it is time to stop eating, our body engages natural satiety mechanisms. Whole foods like a rib-eye steak are delicious and provide long-term energy and plenty of nutrients for fat storage. But could you eat a giant 50-ounce steak at one sitting? No way! Once you're full, you can't eat anymore. The protein and fat in a huge steak activate those natural powerful satiety signals to stop us from eating too much. Even naturally sweet foods like fruit have properties that trigger satiety mechanisms in our body that make it difficult for us to become addicted to them. When was the last time you heard of someone addicted to apples? Or totally, completely hooked on carrots? *Never.*

Few of us are running from predators or hunting game for survival anymore. Instead, we're at the grocery store, scanning the frozen food section, or we're sitting in front of the television with a bowl of chips in front of us. Processed foods like sugary beverages, sweets, chips, crackers, and white bread are *everywhere*, and they're what we, as a modern society, have become addicted to.

When foods are processed in factories, they're stripped of many of their naturally occurring nutrients, including protein, fat, and fiber. This allows the bypass of your body's natural satiety mechanisms. Without fat and protein, the satiety hormones peptide YY and cholecystokinin aren't activated, so they can't signal when we're full. Without the bulk that fiber provides, our stomach's stretch receptors don't know to respond. Essentially, all that's left behind are refined carbohydrates (glucose) and sugar, which cause your pancreas to secrete a surge of insulin and store that sugar as fat. Your blood sugar drops dramatically, and your body craves more sugar. The cycle repeats.

In addition to bypassing natural satiety mechanisms, a food must be highly rewarding to be addictive. Our brains register all pleasures in a similar way, whether they're from a psychoactive drug, a monetary reward, a sexual encounter, or a satisfying meal. Pleasure has a very distinct signature: the release of the neurotransmitter dopamine in the nucleus accumbens, a cluster of nerve cells underneath the cerebral cortex. This area is referred to as the brain's pleasure center.

Illicit drugs like heroin cause a particularly powerful surge of dopamine in the brain. Sugar does exactly the same thing. The hippocampus, the portion of the brain responsible for the formation of new memories, stores this rapid sense of satisfaction, leading you to dream of candy, cookies, and soda.

Our modern processed food addiction is no accident. The food industry spends billions of dollars in research and development to determine the precise combination of salt, sugar, fat, and artificial flavors to maximize temptation, then they cook it up in a lab and roll it out onto conveyor belts, shipping it perfectly packaged to a store near you.

My clients often tearfully confess that they'd rather be addicted to alcohol or drugs than to food. Why? Because they'd get more understanding and empathy from friends and family. When a recovering alcoholic celebrates, you don't buy them a beer. When a drug addict mourns, you don't commiserate with a shot of heroin. But what do people tell sugar addicts to do? Celebrate with sugar. Food is your best friend because it makes you feel good.

Other clients report that their addiction to food is even harder to break than other addictions because food is *everywhere*. For example, I worked with a sixty-two-year-old woman who was a recovering drug addict and alcoholic. In her twenties, she went to rehab twice and was successful in her recovery with outpatient therapy. She says that never in her history of addiction did she struggle like she has with sugar. The reason is that when she goes into a coffee shop, there's no heroin, but there are muffins and sweet rolls next to the cash register. When she goes to church, there's no cocaine afterward at the reception, but there's plenty of cookies and cake. She can go visit her mother without being pressured to have a glass of wine, but she can't get her to understand that she doesn't want a piece of pumpkin pie at Thanksgiving because it's killing her. All these are reasons why she became 160 pounds overweight and became a type 2 diabetic who was placed on insulin.

Because of her long-term history with addiction, she understood and recognized that she was a sugar and food addict, and she was determined to rehabilitate herself for her health, well-being, and family. She changed her diet, but after falling off the wagon and bingeing more times than she can count, she realized she can never, ever have just one potato chip or one slice of cake. Today, she lives a life without the foods that triggered her, and she's reversed her diabetes and lost 160 pounds.

The Importance of Breaking Food Addiction

At this point in my career, I've worked with thousands of clients who are looking to make healthy changes and lose weight. So many of them have told me that part of their struggle to make better food choices is the result of family, friends, and colleagues pressuring them to eat foods they're trying to avoid, especially sugary foods. Their spouses will say things like "I can't believe you won't even try this cake that I went to the trouble of baking for you" during birthday celebrations. And everyone has that one friend or relative who says, "Oh, it's Christmas time, it's okay to have a few cookies!" I can sympathize because I, too, feel this type of pressure to eat to please others. My mother thinks that every time she sees me is grounds for celebration with things like pasta, bread, pretzels, and popcorn—all my old favorites from my addict days.

Most people don't understand just how dangerous sugar addiction can be. Again and again, the research shows that sugar is directly linked to obesity, cardiovascular disease, diabetes, cancer, and many other chronic diseases. Yet, our brains have become hardwired to associate processed foods with instant pleasure, happiness, and belonging. The thought of giving up these foods— even for a short fast—feels impossible.

Eliminating sugar may just tell you more about yourself and your relationships than a decade of therapy. I didn't realize this myself until I had made my own significant transformation. But, in fact, my whole childhood was shaped by my idea of the brain's pleasure response to food. My dad usually came home with his face beet red from the day's stresses, his eyes full of anger. He would avoid us kids for the first hour or so, not wanting to take out his frustrations on us. My mom would quietly prepare a glass of chocolate milk, a plate of pasta, and a bag of potato chips. After she gave them to my father, he would sit and eat them in silence.

When my dad was done eating his big plate of comforting carbs, he'd be eager to hear about our day. As a small child, I could see it clearly: Yummy food made Daddy better. Refined, highly processed food was my family's friend.

Top Ten Addictive Foods

In 2015, a group of researchers set out to identify and rank the most addictive foods. Interviewing five hundred subjects of all ages, they asked them to list which foods triggered addiction-like biological and behavioral responses. Not surprisingly, nine out of ten are almost entirely made up of processed ingredients!

1. Pizza
2. Chocolate
3. Chips
4. Cookies
5. Ice cream
6. French fries
7. Cheeseburger
8. Non-diet soda
9. Cake
10. Cheese

• • •

Breaking Food Addiction
and Unhealthy Habits

Habits are formed by three things: cues, routines, and specific rewards. A cue, such as stress, triggers a routine that leads to a reward—a feeling of relaxation or happiness. For example, you might hate visiting your in-laws, so every time you have to go see them, you hang around the kitchen, nibbling on sugary and salty snack foods. The highly processed foods activate your brain's pleasure center, and your brain is flooded by a feeling of happiness that overwhelms the stress of being in a house you'd prefer to run from. The routine (in this case, eating) is the all-important intermediary between the cue and the reward. If you cut out that routine—or do something else instead—you can avoid the perceived reward.

There are all kinds of cues. There are positive ones like getting married, being on a wonderful vacation, or getting your dream job. There are also negative ones like work stress, sadness, loneliness, and sickness. But no matter the cues, the routine always stays the same: highly processed and refined carbs. Let's celebrate—with ice cream! I'm feeling sad—let's get some ice cream! To change our habits, we must change our routine. The reward—to feel good—does not require the use of highly processed carbohydrates or sweets.

I spend most of my days counseling clients on other ways they can experience relaxation and happiness—that is, feel a *real* reward—without having to depend on food. We use two strategies to help them beat this cycle. First, we substitute fat for sugar each time we have a craving. The fat signals to our brain that we are nice and satiated, and to turn off our appetite. Second, we fast.

Fasting similarly helps regulate our hormones so we regain control over our appetite, but it also gives us freedom. And freedom is the number one benefit of fasting that clients report to me.

I once worked with a fifty-two-year-old executive named Peter. Peter grew up in an Italian family and lived at home throughout most of his education. If he had a bad day, his mom would make him his favorite lasagna and cake. If he had a good day, she'd do the same to celebrate. This situation compares to most of my clients—as well as all their families. Peter's attachment to food was not an addiction but rather a deeply ingrained habit. Together, we worked to find other ways to celebrate his successes and destress from his tough moments, and after about three months of planning, trial, and error, he got into a good pattern. Instead of cake and pasta, he'd reward himself with a yummy rib eye, or some bacon and eggs cooked in bacon fat. He experimented with going to the driving range, boxing classes, meditation retreats, and tai chi until he found a few things that calmed him down. Rather than stuffing his face, he would treat himself to a new golf club to celebrate a success or make time to give his daughter a call to share good news.

If I'm having a stressful workday and I'm fasting, I don't need to eat lunch. Instead, I can take a walk during my lunch hour. This walk will help reduce my stress level and boost my mood and energy. It will also provide me with access to vitamin D from sunlight, which I won't get if I'm stuck in the lunchroom of my office. After lunch I go back to work with a clear headspace, feeling rested and recharged after taking time to do something for myself that has a positive impact on my physical and mental well-being. Fasting gives me freedom to do things for myself and to reconnect with others.

Other Routines That Will Reward You

- Getting a massage
- Taking a nice Epsom salt bath with lavender oil
- Working out
- Getting a manicure or pedicure
- Going out to the movies
- Meeting a friend for coffee
- Calling a friend
- Listening to a new podcast
- Reading a book
- Getting outside for a walk or bike ride
- Tackling a nagging task you've been avoiding
- Meditating
- Writing a gratitude list

You don't need to eat in order to feel close to your spouse or family, and you don't need to snack to cope with an unpleasant task. My husband and I will hike, play a board game, or do a puzzle together instead of eating while trying to connect. Come up with your own routine to replace eating junk.

Preparation is key for success. I ask all my clients to write down their cues and desired rewards, and then come up with a list of routines they can do that aren't food related. For example, a person might say their cue is arriving home at the end of a stressful workday, and their desired reward is food. Instead of eating, I might suggest they go outside and take a walk. This redirected reward is the first stop on their road map to success. Come up with your own list of routines, and keep it in your wallet, purse, or briefcase. Post it on your fridge. Just remember: success is no accident. You can achieve it if you have a *plan*.

So, now that you understand the science behind fasting, what foods to eat and which to avoid, and how to think differently about your relationship with food, let's get in the right mindset and start to formulate that plan. *It's time to get in the fasting lane!*

PART II

Prepare to Fast

CHAPTER 7

Ready, Set, Goal!

EVE MAYER

You are at the beginning of an exciting journey. You may be tense and eager, ready for the gun to fire so you can launch toward the challenge. Or—if you're like most of us who've battled weight and health challenges for years—you might be anxious about what lies ahead.

Desperation and self-doubt are some of my most familiar companions. They sit next to me, invisible to the rest of the world, whispering words of discouragement softly into my ear. They remind me that I have failed at eating better and getting healthier for most of my adult life. They snicker at my hope and laugh at my thoughts that this time might be different. Self-doubt nags at me, asking: *Why would I ever think I can fast if I haven't been able to stop shoveling massive amounts of food into my mouth for decades?*

It's time to stop thinking this way. No matter how many calorie-restriction diet plans you've tried in the past . . . yes, you can do this.

Remember: you are not the problem. You've simply been fed incorrect information for years. But with the right knowledge, your body can heal itself. In fact, it may be even easier than you think because it's possible your body and mind are not as broken as you suspect. Like me, you may discover that beneath the extra weight, you are healthier than you could ever imagine. This is why a plethora of people who fast heal their type 2 diabetes, resolve their high blood pressure, or go off their medications with shocking speed. My father is a case in point.

When my dad saw the success that my husband and I had with fasting, he decided to give it a try as well. He reduced his sugar intake (although my mom told me he was still eating ice cream or cookies at least once a week) and cut down on breads, pasta, and potatoes. He followed an easy fast, skipping breakfast most days and reducing his snacking, and he lost fifteen pounds the first month.

About three weeks in, Dad felt so dizzy that he had to stop driving. He went to his doctor, reported his dizziness, and said he'd been fasting. He then explained to the doctor that he had lost weight, and that while his blood pressure had improved, he was still taking his prescribed blood pressure medicine.

The doctor told him to stop fasting immediately but to continue eating healthy, reduced portions. The doctor also felt that it was very possible he might have vertigo, so he prescribed another medicine and suggested that my dad see a specialist to help resolve it.

My mom was worried about my dad, but she was skeptical of the doctor's diagnosis of vertigo. It seemed suspicious that her husband had suddenly developed vertigo for the first time in his life. She felt that it was more likely that the fasting and changes in diet had to do with his dizziness. It turned out she was correct. Because of his fasting and low-carb diet, my dad's blood pressure had gone down *fast*, in three weeks. That meant he was currently taking more medicine than he needed and was becoming dizzy because of it. My dad cut his blood pressure medication dosage in half and continued with his new way of eating and fasting. Over the next week, his dizziness was gone, and he decided to cancel the appointment with the vertigo specialist. He also left the vertigo pills unused in the cabinet.

In a few more weeks, after carefully monitoring his blood pressure three times a day and losing a few more pounds, he was able to cancel his blood pressure prescription completely, as well as to get down to 229 pounds for the first time in twenty years.

How did he accomplish this huge, life-changing feat? Because he'd set a powerful goal from the very beginning. My dad loves to ski more than anything in the world, yet he hadn't hit the slopes for many years due to his high blood pressure and extra weight. He told himself that if he could get his weight into the 220s, he would go skiing. Now he's planning his trip!

Lies We've All Heard About Fasting

- **FASTING WILL MAKE YOU SICK.** Quite the opposite! Fasting can lower your risk of heart disease, cancer, type 2 diabetes, and high blood pressure.
- **FASTING WILL CAUSE YOUR BLOOD SUGAR TO CRASH.** Your body does a wonderful job regulating blood sugar levels, so there's little chance of your body having a negative hypoglycemic response.
- **FASTING WILL SLOW DOWN YOUR METABOLISM.** There is no research that shows that fasting—even fasts up to three days— suppresses metabolic rates.
- **YOU'LL DIE OF HUNGER.** This is my favorite! Have you ever skipped a meal? Look what happened! You didn't die. I know this because you are reading this book.

Goals Start with You

If you are out of the habit of goal setting, it is time to get back in the groove.

Many people feel comfortable putting everyone else in their lives first when it comes to setting goals. They make a goal that their sixth graders will get on the honor roll this semester, or that their spouse will get a raise by the end of the year. When did we get so busy helping others that we forgot to take a moment to consider what *we* truly desire?

Now is the time for *you*. You are reading this book to discover how your body, mind, and life can be changed when you incorporate fasting. Start by deciding what you want to gain by fasting. The answer can be anything at all.

When I began this journey, I had only one goal. I wanted to be hot again!

Before I started fasting in my mid-forties, the last time I felt good about my body was when I was eighteen. It was *beyond* time for me to wear a two-piece bathing suit again. Was I vain? Who cares? I was doing this for me, and my vanity drove me toward my goal. After a few months, I got down to 195 pounds (a thirty-pound weight loss), and I wore that two-piece bathing suit. I felt better about myself than I had in more than two decades.

Your goal is individual to your wants and needs—and it is vitally important to your success. Reaching your goal may not be fast or stress-free, but nothing worthwhile ever comes easily. Picture your fasting skills as a muscle you must exercise, rest, and grow. Some days you'll flex that muscle with ease. Other days are going to be a challenge. It's during those times that focusing on your goal becomes important. When you've had a few difficult days in a row and find yourself wondering, *Is this worth it?*, it's your goal that will remind you why you are on this journey—and that *you* are worth it.

Goals Change

Over the course of my journey, my goals shifted. As I gained more confidence with fasting, learned new techniques, and lost more weight, I began recognizing how much easier it was to keep up with my husband and my eleven-year-old daughter. For years, I'd suffered from allergies, recurring upper respiratory infections, and bronchitis, but I stopped getting sick. I didn't need to constantly take medicine for my respiratory issues. I stopped getting headaches every other day. My teeth also got healthier—much to the shock of my dentist.

Feeling so much better and seeing numbers on the scale that I hadn't seen since I was a teenager—and with less work than I'd ever put into weight loss before—was a bewildering experience. But I realized I was on to something, so I set an even more challenging goal. Knowing that body composition (the percentage of body fat in relation to muscle) is as critically important to good health as weight, I decided that I wanted to reduce my body fat percentage by 5 percent.

This road can get bumpy sometimes, so I want you to create one very specific goal—or, if you must, no more than two—that will help keep you motivated. I'm sure you want to achieve many things, but limiting your focus to the one or two most important goals will help you keep your eye on the prize and lessen your chances of getting overwhelmed. The amazing thing is that once you accomplish just a few small goals, you'll find that some of your other hopes start to come to fruition, too. For example, maybe your goal was to go from an A1c level of 7 percent (diabetic) to 6 percent (prediabetic). One of the unexpected side benefits could be that the tingling in your toes—which had been a constant nuisance—stops. Or that, suddenly, you're able to walk up a flight of stairs without losing your breath.

Your goals should be clear, specific, concrete. For example, an unclear goal would be "I want to be more active" (active how and when?) while a clear goal could be "I want to walk a 5k."

Goals People Have Shared with Me Before Fasting

- Get off type 2 diabetes medicine
- Wear a new dress to my high school reunion
- Complete a triathlon
- Reduce my body fat percentage
- Buy pants that aren't stretchy
- Get pregnant
- Get sick less often
- Get off blood pressure medicine
- Wear high heels that don't come in a wide size
- Be able to walk with the dog one mile
- Stop having migraines
- Have better focus at work
- Play with my grandkids
- Lose a certain number of pounds

Once you're clear about what your goals are, I want you to write them down in three places. You can keep your goals on a piece of paper in your top desk drawer at work, scribble them on your mirror at home, type them into your phone, or make them your computer screensaver. Every time you come across your list, read it out loud three times, adding the words "I will" at the beginning of each goal. You can also say them silently in your head.

So often, we put aside what we want for "just one more day" until the "one more day" stretches into months and years. Keeping your goals in the front of your mind and reminding yourself of them daily will help reinforce them.

MEGAN RAMOS

You can't just name your goals; you also need to create a plan to achieve them, and then follow through with your plan. So, what makes us get up and take action to make our dreams a reality? Motivation. It is the force that moves us from being stuck or stagnant to doing something proactive. As a health educator, I'm here to explain how to do that.

What Motivates You?

There are two types of motivation: intrinsic and extrinsic. Intrinsic motivation means that what drives a person comes from within, and that their desire to perform a specific task is in accordance with their belief system. Eve's desire to "get hot" and "feel hot" is an example of intrinsic motivation. Other examples would be to feel better, have more energy, and reduce your risk of type 2 diabetes.

In extrinsic motivation, an individual's stimulus is external, even though the results will still benefit the individual. A classic example of extrinsic motivation is money. In this case, that might mean reducing healthcare costs or becoming more focused and efficient at work to achieve a promotion. Goals and the motivations behind them are as relative and dynamic as the people who set them, so it doesn't matter if your objective is extrinsic or intrinsic. What matters is that the goal is powerful to you.

As an educator, one of my primary intentions with each client is to identify what motivates them. Are they sick and tired of feeling sick and tired? Are they struggling to pay for their medications? Or do they want to simply feel better in their own skin? I spend time getting to know everyone I counsel and figure out what makes them tick. Once I understand that, I use it as leverage to help inspire them through their journey.

Sometimes I meet a client whose motivation isn't in alignment with their lifestyle. For example, one woman told me she wanted to lose weight so she could enjoy more time with her family. Unfortunately, she said the only time she had to connect with her family was during mealtimes—and she was going to be skipping meals! I worked with her to find other opportunities to spend time with her family, and, after a few months, her fasting habits stuck.

Overcoming Bad Habits and Bad Times

Our motivation for weight-loss success can often be derailed by bad habits or life's unexpected curveballs. When things are stable, we're able to fast successfully and be mindful of our sugar intake. When life is chaotic, we tend to fall back into our deeply ingrained old dietary habits.

This is why it's so important for me to understand what is motivating a client to make a change. If someone is trying to improve their health so they can be around to see their grandchildren grow up, for example, that motivation is a great way for me to help them stay inspired or get back on track. If that client reports having a lot of energy while they were babysitting their grandkids, I remind them of their motivation and help them create positive associations with fasting and their goal.

I use two tools to help motivate me. The first motivating tool is a photo of myself wearing a bikini while hanging out with my friends in Miami Beach on my twenty-first birthday. I carry this picture around because—as much as I say my goal is to be happy and healthy—like Eve, I also want to look good. Am I ashamed of that? Not at all! I also carry around the blood test results that diagnosed me with type 2 diabetes. This piece of paper scares the hell out of me, and I've used it dozens of times over the last nine years to prevent myself from eating an entire large pizza alone after a bad day.

Easy Ways to Motivate Yourself
- Go online and read testimonials from people who have reached their weight-loss and health goals through fasting.
- Listen to relevant podcasts and save the most motivating ones for review when you need an extra kick.
- Read books about fasting and write down pages that inspire you for future reference.
- Carry around a note to yourself about why you're doing this, so you can read it when you're tempted to break your fast on a stressful day.
- Keep old blood test results with you to inspire you to stay the course.

When You Lose Sight of Your Goals

Years ago, I was diagnosed with a very, very early stage of cervical cancer. Fortunately, I required minimal intervention, and three months later it was like a bad dream. But, at the time, it was very real and deeply frightening. When I get to a place where my usual goals aren't providing sufficient motivation, I recall the feelings I had during this dark time in my life, and I think about how many cancers are associated with obesity.

After I started sharing this strategy with clients, I was surprised at how many of them were also motivated by fear. Maybe they're not concerned about a cancer diagnosis, but they might worry about a trip to the hospital because of mild chest pains, a diagnosis of prediabetes, or a hereditary family condition. For example, a client named Rose once came to us after a terrible medical scare involving her hip. At forty-nine years old, Rose was 5 foot 3 and 180 pounds. She underwent a total hip replacement, and during her long, painful recovery she put on weight, topping out at 201 pounds. Four months after her surgery, Rose was stretching, and her hip socket came out of joint. A specialist was able to repair it, but he said it was a terribly difficult process. "Why?" she asked. He responded, bluntly, "Well, you're short. But also because of the weight."

Rose *never* wants her hip to be a problem again, so she's now following a keto diet and doing 36-hour fasts when she can. She's down to 160 pounds and wants to lose 20 to 25 more. Best of all, she says she's had no further hip problems, and her orthopedic surgeon is thrilled!

Another tool I offer clients when they're losing sight of their goals is visualization. You can visualize positive events, like what the outcome of your goals would look like in real life (for instance, spending more time with family or lying on the beach in a two-piece bathing suit). You can also visualize negative outcomes, like your doctor delivering the news that you've been diagnosed with type 2 diabetes. Whether you're imagining happy or sad experiences, visualization engages your senses and your emotions in a powerful way that can help you keep focused on your goals.

When I look at my old blood test results, I don't just look at the numbers for motivation. I try to recall how I *felt* when I got those results. What did my doctor's face look like when she gave me the bad news? How did it feel to hear the disappointment in her voice? My body still remembers the chills down my spine, the tears forming in my eyes. That's enough for me to push through a fast when I'm tempted to eat a bag of pretzels.

Prioritize

One of the biggest struggles I have with clients is that they often have too many goals, and because of that, they want to dive into a fasting program before they've had a chance to prepare for it. For example, they may want to lose 150 pounds, stop taking their medications, reverse their type 2 diabetes, *and* avoid the Alzheimer's diagnosis that runs in their family. These are wonderful goals, but they can't all be achieved at once, no matter how hard you try.

I help clients prioritize their goals by asking two questions: "What is going to kill you first?" and "How strong is your fasting muscle?" Most are aware that they need to lose weight in order to reverse their type 2 diabetes, which will result in a reduction in the number of medications they take every day. If they don't beat that, they're putting themselves at a high risk for metabolic-related cancers and Alzheimer's disease. Therefore, diabetes is the disease that may kill them first, and it's what they need to focus on first and foremost.

But these eager clients can't and shouldn't dive into something as intense as an all-day fast. Many of them haven't fasted longer than twelve hours for a blood test before, so fasting for twenty-four hours straight, right off the bat, isn't going to be an easy task. With priorities, you can focus, which may help an impatient person proceed into fasting slowly. For example, you may choose to skip breakfast two nonconsecutive days a week, then increase the frequency bit by bit.

As Eve said, I encourage you to write down a list of your goals and prioritize them. The most successful people concentrate on one thing at a time, and they know that trying to do too much at once will inevitably lead to failure. You don't want to take chances on your health, so be patient and consistent. If you are, you *will* reach your goals.

CHAPTER 8

Get Your House in Shape and Your Family on Board

EVE MAYER

Your goal is firmly in place, and you've decided what types of food fuel you and what foods might derail you. How can you give yourself the best chance for success? Some of the answers lie right in the comfort of your home.

Cleaning House

It is time to take stock of the spaces you live in and do a clean sweep. Examine your fridge, your cupboards, as well as your car or your desk drawers at work—any place that you spend more than one hour a week. You want to get rid of the sugary and processed foods you've decided *not* to eat most of the time—those tempting, pesky little items so easily accessible that they'll be magically in your mouth before you realize it.

Start with your kitchen. Identify all the foods you've decided to refrain from eating, then get rid of them within twenty-four hours. Donate food items that are acceptable to a food bank and give others to family members or friends. And I know this may make your food-loving heart wilt, but if you cannot donate the food in twenty-four hours, I want you to throw it away. Why? Because the longer food you have decided is harmful to you stays in your home, the better the chance it will end up in your mouth, and you will put off your goals for one more day, one more week, or one more month.

After you've cleaned out your kitchen and pantry, you need to repeat the process in the other rooms of your home. See that candy dish? Get rid of it. Replace it with flowers or a keepsake you enjoy. Next, it's time to examine the contents of your purse, your backpack, your storage area, your car, and your garage. Clean out all the items you were saving for a rainy day. You might be thinking, *I'm giving up ice cream* most *of the time, but I still plan to have it* sometimes. *Why should I throw it out if it's far away, in the deep freezer in my garage?* Your goal is to make the foods you only want occasionally harder to get. Eating a gallon of ice cream that took you five minutes to retrieve from the garage is easier than driving fifteen minutes to the ice cream store for one scoop. The extra work forces you to examine if you *really* need that ice cream.

Living and Working with Others

If you live with others who are not eating the same way as you, you likely won't be able to get rid of their food. However, there are ways to make things a bit easier on yourself. Separate your food from theirs—in the fridge, pantry, and elsewhere—making it clear who the items belong to. If you have the opportunity to cover their food with foil or place it in snack boxes, do so. Out of sight, out of mind!

You want to keep your supply of foods front and center. In the fridge, store veggies in a decorative bowl or on plates you find pleasing to the eye. In the pantry, organize your food neatly, making all healthy choices visible in glass containers. Put labels with your name on the packaging that contains your food, and if you are an overachiever, place a whiteboard in the kitchen or use your phone to create a menu plan for the next few days or week.

Next, let's examine your workplace. If you work from home, then you already know what to do. Open your desk drawers and give away those chips, granola bars, and candy bars. In fact, do away with *all* the food items in your office. If you work outside the house, go to the shared work refrigerator and throw out the old food you haven't yet brought home. Do the same at any location where you might have stored snacks for yourself. Do you always grab pretzels from the break room on your way to the bathroom? Try taking a different route when nature calls. In the next section, I'll explain the importance of skipping snacks and show you how to go about doing it, but, for now, know that paving the way for new habits is essential to your success with fasting.

Making Your Family Part of Your Team

If you're single or living alone, without a roommate, kids, or partner: congratulations, you get to skip this step!

I still remember those blissful single, childless days, stretching out the width of the bed, sleeping till ten on weekends, and having the TV remote all to myself! But then things happened. Marriage, a daughter, a divorce, falling in love, marriage again, and then a puppy. Many of us are living with a roommate, spouse, friend, parent, children, or some combination of people sharing our lives on a day-to-day basis.

The people you make your home with almost always affect most areas of your life, and they can have a huge influence on when and what you eat. Now that you have decided to implement some changes to your diet, it's important to communicate your plan to them, ask for their support, and in some cases suggest that they join you on this journey.

It's possible your partner already has some of the behaviors that you wish to adopt. Perhaps they eat healthy foods, don't snack, or even practice intermittent fasting themselves. If this is the case, you will likely find supportive smiles coming your way when you share your plans for changing your eating.

Some families live together yet rarely eat together because of varying life, school, or other schedules that make it nearly impossible. If this is your situation, there's not much of a need for getting buy-in from them. But if you eat at least one meal each day with the people you live with, they'll need to be consulted. While your decision to fast is a very personal one—and you have every right to make it—problems or differing opinions may crop up, and it's best to prepare yourself for them.

The cornerstone of your relationships is usually your partner, so it's probably best to discuss your new eating plans with him or her first. Just remember that it doesn't have to be dramatic. When you start, you are simply testing the waters on fasting, so there is no need to proclaim to your partner that you'll be fasting every other day for the rest of your life. Instead, start slowly and gently, just as you will with fasting. Tell your partner about your goal, how you plan to reach it, and what type of support you specifically need from him or her. Here are two examples:

Frank, I'd like to lower my blood sugar to healthier levels, so I'm going to snack less between meals. I plan on buying fewer snacks at the store. If there are specific things you want me to get for your snacks, please let me know.

Susie, I'm trying to lose weight, so I'm going to skip breakfast on weekdays. I'd still love to have coffee with you in the morning, but I'm going to get ready for work right after that. Will you please pardon me from breakfast during the weekdays for a while?

Some partners will be happy to cheer you on on your new eating regimen. Other might have feedback or want more information. Listen to your partner's concerns, take them seriously, and answer them completely. If they ask you a question about food or fasting you can't answer, tell them you aren't sure at that moment, but you will get the information to them soon. Then do your research and share what you've found with your partner. Our website, FastingLane.com, is a useful and instructive place to start, with blogs, articles, and a podcast. You may need to research again and again! I'm almost two years into my new way of life, and I'm still learning new things every few days.

It is understandable for a partner who cares about you to be worried. I can't imagine what my response would have been a few years ago if mine had told me he was planning to skip meals. I would have been worried about his mind-set and his health, and I would have warned him that his metabolism was going to slow down (it won't!). If your partner questions, prods, and explores—just like I would have done—try not to get frustrated. Listen to their questions and be happy that someone cares enough to show concern for you. Use it as a learning opportunity to find the answers you haven't explored yet.

Next, think about how these changes will affect your partner and address those concerns. Perhaps you're responsible for cooking for the family, but you want to skip eating dinner on some nights. It is reasonable for your partner to be concerned about how that might affect the rest of the family. Try to anticipate this and address it before you're asked, then come up with creative solutions that are helpful to both parties. Here is an example:

> *I'm going to be skipping dinner most Tuesdays for a while. On Mondays, I'm going to cook enough food to have leftovers for Tuesday's dinner. Could you please eat dinner with the kids on Tuesdays while I take the dog for a run?*

I expect your partner to ask questions, show concern, disagree with some of your methods, be curious, and even sometimes buck at the changes you're proposing. Sometimes you strike gold, and your partner decides to join you. This will certainly make life easier, but if they don't, ease them in by inviting them to taste the new, healthier dishes you plan to prepare. Ask them to join you at the gym or even to skip breakfast once. Your partner has the right to decline your offer, and it's important to respect their decision. I understand how frustrating this can be if you live with someone who's overweight or suffering from weight-related health conditions. You're probably thinking, *The solution is right in front of you!* But we can't force anyone, including our partners, into decisions they haven't chosen for themselves.

You also deserve kindness from your partner through the

changes you are making. It is okay for them to question you, but it is not okay for them to interfere with your progress. If your partner is unsupportive through this process, I would suggest communicating your needs very clearly. You might say something like, *I'm making these changes to improve my health. Your love and kindness would be a huge help.*

During any time of transition, it's important to be patient with the people around us. We made the decision to change, not them. We are all so quick to be critical of our partners, and often, I find this is because we have become habitually critical of ourselves. It is so easy to treat our partners with the same cruel voice we hear inside our own head. So, speak kindly to your loved ones, and you may find it easier to speak kindly to yourself.

MEGAN RAMOS

Many couples decide to fast together, which can be beneficial if both parties stick to their plans, don't sneak food, and remain emotionally supportive of each other. However, if you're in a male/female partnership and decide to embark upon a fasting lifestyle together, you need to be mindful that your patterns of weight loss may be very different.

Fasting Differences Between Men and Women

Most of the women I've worked with have tried every diet out there. Whether it's Jenny Craig, Weight Watchers, or any number of juice cleanses, almost all these weight-loss plans have one thing in common: they focus on reducing calories and increasing energy expenditure. Not only are these ladies demoralized by their continued failure, but their BMRs are at rock bottom. Constant dieting has killed their metabolism.

It's quite the opposite for men I see for the first time. These men may be carrying twenty, fifty, or one hundred extra pounds, but they haven't tried many diets. They've either ignored their weight gain or been in denial about it, so, unlike their girlfriends or wives, they haven't obsessed about trying to lose the extra weight. Because of that, their metabolism is still high.

Women are more hormonally complex than men, which makes balancing hormones—a process fasting kick-starts—somewhat tricky. Couple that with vastly different metabolism levels, and men and women who try to fast together can find themselves losing weight at very different rates.

Jason and I have observed the following patterns when it comes to men and women who start to fast:

- **WEEK 1 OF FASTING:** Men can expect to lose a half pound of body fat for their first 36-hour fast. Women will lose about a quarter pound.
- **WEEKS 2–4 OF FASTING:** Men lose about one pound of body fat per fasting day (i.e., a 36-hour fast).
- **WEEKS 4–6:** Women's metabolism catches up to where men are, and they begin to lose about a pound per fasting day. Men have leveled off, and they lose about a half pound of body fat per fasting day.
- **BEYOND WEEK 6:** Both men and women lose about a half pound of body fat per fasting day.

Typically, men lose more weight with 24-hour fasts, while women may need to do 36-hour fasts to achieve the same degree of weight loss. This may be frustrating to women, but if you manage your expectations as a couple, knowing that each of you can and will reach your goals differently, you can avoid getting upset. Even though men and women lose weight differently, couples often find that dieting and fasting together is easier than apart. Just remember to educate yourself, and always communicate.

JASON FUNG

There are several other hormonal changes that occur during fasting, and one of them typically impacts women more than men. I'm talking about the increase of human growth hormone, or HGH.

During a fasted state, the body produces higher levels of HGH, noradrenaline, and cortisol. These three hormones are referred to as the counter-regulatory hormones, and they help increase blood glucose at times when it can't be derived from food. HGH, which is produced by the pituitary gland and secreted during sleep, is central to healthy growth in children, but it also helps maintain muscles and bone mass in adults. When an adult doesn't have enough of it in their system, they may develop more body fat and lose bone mass and muscle.

HGH secretions go *way* up during fasting. In fact, according to a 1988 study, a two-day fast can help you produce five times as much HGH! This is greatly beneficial for men *and* women because strong, lean, sturdy bodies are better for health than leaner, weaker muscles and skeletal frames. However, a pound of muscle is denser than a pound of fat, so many women are shocked when they step on the scales after a fast and discover that, while their pants fit better, they haven't lost any weight.

I've found that men are less concerned with this issue. They tend not to step on the scale as much, so they don't care as much about weight. Not so for the women in their households, who tend to feel greater disappointment.

To the women out there reading this book: you don't need to feel concerned. As Megan explained, over time, women tend to lose weight at an equal pace as the men in their lives. Thanks to HGH, stronger, sturdier muscles and bone mass are the icing on the cake, giving you a healthier body, not just a slimmer one.

CHAPTER 9

Sex, Pregnancy, and Fasting

MEGAN RAMOS

If digestion, weight gain, fat storage, fat burning, muscle growth, and bone mass are, in part, ruled by hormones, what about *those* hormones? You know what I'm referring to: the hormones you feel pumping through your veins when you're with someone you're attracted to, that then drive you to reproduce successfully? As you embark on a lifestyle of fasting, should you be ready to put your sex life on hold or postpone your dreams of getting pregnant? Not at all. If anything, you should prepare for a re-energized libido. And, while we don't recommend fasting during pregnancy, it can help pave the way to getting pregnant.

• • •

Fasting and Sex Hormones

Some women worry that fasting will deplete their energy, which will cause their sex drive to plummet. The opposite is often true. Because fasting helps regulate all hormones, women find that fasting actually *increases* their sex drive. Fasting also may boost vaginal moisture, which helps many women enjoy sex more than they did before.

Don't expect these changes to happen overnight, though. Positive sexual side effects usually occur within the first three months of consistent fasting—whether that's skipping breakfast every day or doing a 36-hour fast twice a week. Consistency is key: a regular practice of fasting helps keep hormone levels stable, and going off it may cause them to fluctuate.

There are some women who experience a decreased sex drive due to fasting, but this is rare and can usually be attributed to inadequate intake of dietary fat (from natural fat sources such as avocados, olive oil, coconut oil, and fatty fish) or sodium on eating days, leading to a nutrient deficiency. Usually sex drive returns or increases once sodium levels are restored, though this can take four to six weeks, depending on compliance and the woman's unique physiology.

Sodium intake is very specific to each individual, so I encourage you to salt food according to your taste on eating days. On fasting days, you can consume sugar-free pickle juice or you can add salt to water or place a pinch on your tongue and then drink water. Whichever way works. You require, on average, between 1 and 3 teaspoons of salt per day, but if you have a certain medical condition like high blood pressure, you may need to avoid salt entirely. Always check with your doctor. And if you've noticed that your salt intake has no effect whatsoever on your libido, you're obviously ingesting the right amount for you!

JASON FUNG

Research into the relationship between sex hormones and fasting is almost nonexistent, so Megan, Eve, and I can only speak from our personal experiences and our work with clients throughout the years.

One happy side benefit of fasting may be a pregnancy that had previously been impossible due to a condition such as polycystic ovary syndrome. High insulin levels increase production of testosterone from the ovaries, which eventually leads to the abnormal growth of multiple cysts. By reducing insulin, fasting may help reverse this condition.

Jennifer—the PCOS sufferer you met in the introduction to this book—proved that fasting can help alleviate PCOS as well as lead to a successful pregnancy. So when a client asks, "I'm trying to get pregnant. Can I still fast?" my answer is always an emphatic "Yes."

By helping you lose weight, fasting may also help prevent some of the unwanted complications that can occur during pregnancy, such as gestational diabetes and high blood pressure. That said, it's important for a woman who is fasting for fertility reasons to closely monitor her cycles and stop fasting as soon as she knows she is pregnant. Pregnancy is a time for growth, and fasting restricts the nutrients that are needed for growth. Fasting may also negatively impact the quality of breast milk, so we recommend that nursing mothers not fast. However, it's perfectly okay to focus on time-restricted eating during this time—that is, eating meals within an eight-hour window and abstaining from snacking.

EVE MAYER

You might be thinking, what is a chapter on sex doing in a fasting book? I understand that it may not be pertinent to everyone, but the truth is that it's a question that comes up a lot with Dr. Fung and Megan's clients, and it's also one that comes up with my very best girlfriends who've been on the fence about fasting. They wanted to know how they'll have enough energy to have sex. They were curious about whether they'd get headaches that would cause them to lose interest—or feel too lousy—to be in the mood. And they wanted to know if fasting would cause their hormones to go into overdrive, causing them to get more emotional or to fight with their partners.

I feel lucky that I am in a committed relationship with a man whom I find to be a stone-cold hottie. I thought fasting would make me grumpy, irritable, and not in the mood. Boy, was I wrong! Fasting gives me more energy. In fact, when I first began fasting for multiple days I had so much energy that I often couldn't get to sleep. Hmmm, what is one to do when they're lying in bed awake at 1:00 a.m. with their husband? I think you can fill in the blanks here.

Getting Pregnant, Girlfriend to Girlfriend

If I knew then what I know now, I absolutely, positively would have fasted when I decided to start trying to become pregnant.

Fifteen years ago, I weighed three hundred pounds, was prediabetic, got sick every month, and had PCOS. I knew I had to get my body in a better condition to healthily carry a child, so I went to doctors to figure out a solution to my obesity. I had already tried every diet and exercise and therapy with dismal results; I just kept getting fatter and unhealthier.

Mostly because I wanted to have a child, I turned to the suggestion of a doctor to explore bariatric surgery. I chose the lap band, and it did help me shed quite a bit of weight. Even with eighty pounds lost, though, I remained prediabetic, was often sick, and still suffered from PCOS, which required me to seek the help of a fertility doctor to get pregnant.

From my own experience, I can say, without question, that I would have used fasting instead of surgery had I known about it and understood it. Because of fasting, at forty-five years old, I am no longer prediabetic, and I no longer suffer from PCOS. My forty-five-year-old self is so much healthier than she ever was at thirty, when I ate nine times a day.

Every week, I read stories online and hear from women who used fasting to improve their health, including their fertility. Of course, it is no magic bullet, and infertility is complicated, with different factors involved for every woman. But it is a strategy that isn't always discussed—and one that I have seen help a lot of women.

CHAPTER 10

Working with Your Doctor

EVE MAYER

For more than twenty-four years, I followed the advice of doctors to overcome my weight problems. They told me what to eat and what not to eat. I listened to what they said so closely that I tried hypnosis, therapy, rehab for binge eating, doctor-supervised weight-loss programs, calorie-restriction diets, daily walking, growth hormone injections, diet pills, and three bariatric surgeries. I always lost weight at first. But I always felt hungry, overwhelmed, and unsatisfied. I always gained the weight back, and I often gained even more.

For decades I experienced respiratory illnesses about every two months. I had frequent bouts of bronchitis and a few cases of pneumonia. I was prediabetic, and I went on medication for it. I struggled with polycystic ovary syndrome and couldn't get pregnant without bariatric surgery and the assistance of a fertility specialist. I was on antibiotics and steroids often. I had recurring headaches at least every other day. My dentist couldn't understand why my teeth seemed to be rotting away, even though I had regular checkups and practiced good dental hygiene. Through it all, I accepted that I was simply a sick person with a compromised immune system.

Then I got better. And I did so when I stopped taking my doctors' advice.

Let me step back a bit and qualify what I just wrote. I know that in order for you to improve your health, you're going to need support from others. Your doctor may be one of them, especially if you have a health condition that requires you to take medication or be under regular care. But questioning your doctor's advice about your diet is okay. Trusting your own gut is even *more* okay.

The truth is that many doctors give terrible advice when it comes to nutrition. During medical school and the years of training after it, they've only received a few hours of nutrition education—as Jason Fung will attest to. Why would you put your food choices solely into the hands of someone who hasn't had much more instruction in that area than you have?

Think about it. You didn't ask your doctor's permission—or seek his or her advice—while you were eating poor-quality foods. When you ate two pieces of cake, did you call your doctor? Every time you ate between meals, ordered a Grande Vanilla Latte with whipped cream, grabbed a second doughnut at work, ordered into the speaker of a fast-food drive-through, supersized something, or had a fourth meal that day, did you ask your doctor's permission first?

No. You most certainly did not.

If you didn't get your doctor's permission to eat an extra meal, why do you need to ask him or her about skipping a meal?

Unless you're under medical orders not to fast, you really don't.

If you *do* decide to speak with your doctor, be prepared to be told that fasting is a fad. My answer to that is: if fasting is a fad, it sure must have some staying power because it's been practiced by humans for *thousands* of years. You should also be ready for your doctor to insist you stop fasting and instead tell you to limit how many calories you eat. This approach has been working well for you so far, right? Nope. So why not stop snacking first and see how you feel? Then skip just one meal to see if this is something you want to pursue? If you find it compelling and want your doctor's advice, then make an appointment. But, I recommend that you go to the doctor only if you need him or her to offer you a specific and important piece of information or service related to your health.

When *Should* I See a Doctor?

There are a few valid reasons to see a doctor when you start to change your eating habits and incorporate fasting into your life. They include:

- You want a medical record of your weight and accountability with a medical professional you trust.
- You have diabetes, high blood pressure, or other diseases that need to be managed during fasting.
- You are on medication that needs to be adjusted because you take it with food, or you have lost weight and need a different dosage.

At your visit, be clear about the reasons you've made the appointment, then state what action you have taken, what the results have been, and what your question is. For example:

"Dr. Garzos, I've been eating three times a day instead of five times a day for two weeks. I've lost 4 pounds and my blood sugar has been closer to normal. How do I know when to adjust my metformin?"

See? That was easy!

I am not a doctor. I am a person who failed for twenty years, and then I succeeded, ending my suffering after following my own heart and my own advice. I don't know for sure that fasting is the answer for you because only *you* can find the right path. I believe that if the other things you have tried failed, there is a strong reason you should try fasting. And I think there's a ridiculously strong chance you may succeed.

JASON FUNG

There are certain things that doctors are great at doing. Prescribing medications? Yes. Doing surgery? Yes. Nutrition and weight loss? Just as Eve wrote, the answer is no. You might be a little stunned to hear that admission, coming from a medical specialist like myself. But it all comes down to a physician's training and what they see as their circle of competence.

Medical training extends more than a decade, and there is barely any attention paid to nutrition or the equally thorny question of how to lose weight. In medical school, the standard curriculum includes a mandated number of hours for nutrition, which varies depending upon where you did your training. Generally, during the four years of medical school, you receive ten to twenty hours of instruction on weight loss, and it's about as insightful as your latest issue of *Cosmopolitan*. Eat less. Move more. Cut out 500 calories per day and you will lose about a pound of fat per week. This is the same old, same old advice that studies have shown *doesn't work*.

When I was doing my medical training at the University of Toronto and the University of California, Los Angeles, nutritional lectures centered on things like the metabolic pathways of vitamin K or learning the pathway of vitamin D activation in the kidney and skin. Yes, perhaps you might consider that a lesson in nutrition, but it's really much closer to biochemistry. We also learned about diseases such as scurvy (vitamin C deficiency, which was common centuries ago among men who spent most of their lives as sea) and pellagra (niacin deficiency). Knowledge of scurvy certainly came in handy during exams, but the last person whom I diagnosed with scurvy was, uh, no one. That's probably because I am a modern-day physician and not a pirate of the Caribbean.

As a doctor, most of my clients want to know things like: Should I eat more carbs? Fewer carbs? More fat? Less fat? Is sugar bad? How often should I eat? How do I lose weight? Most medical schools provide less training in these real-life nutritional issues than most health clubs or gyms. Sure, every medical student knows that obesity plays a dominant role in metabolic diseases like type 2 diabetes and metabolic syndrome and, in turn, these metabolic diseases raise the risk of heart disease, stroke, cancer, kidney disease, blindness, and amputations. But those are *diseases*, and doctors are trained to treat diseases with drugs and surgery rather than providing solutions that address their root causes.

Does it change after medical school? Yes, it gets worse. A doctor's specialty training, internship, residency, and fellowship last another five years after medical school, and during that time there is no formal curriculum for nutrition education. That's another five years where doctors learn that weight loss has nothing to do with them. Leave it up to Weight Watchers, Jenny Craig, and magazines. It's not real medicine.

So, should you talk to your doctor about weight loss? Would you ask your plumber to remove your wisdom teeth? Would you ask your barista to check your vision? No. It's *possible* your doctor is an expert on nutrition and weight loss, but it's more likely that he or she is not. Unless you are currently under the medical care of a doctor for a specific ailment, please trust yourself to make wise decisions about the food you eat. After all, it's *your* health.

CHAPTER 11

Letting Go of Shame

EVE MAYER

The first time I ever squeezed my body into a pair of Spanx, I was in Paris.

Paris is a strange place for a plus-size woman. I was there to speak at a conference, and the women in the audience were young, sophisticated, and, most of all, thin as rails. These gorgeous creatures looked beautiful and healthy, and I . . . well, I did not. With my 240-pound body stuffed into a pair of Spanx, I felt like a giant sausage in a tight casing.

I'd heard women from around the world praise Spanx, saying they could smooth out your extra curves while simultaneously obliterating your body image issues. I couldn't help but admire the company's founder and CEO, who had the guts to make a business that would help women feel and look their best. But, too embarrassed to try on my Spanx in the store's fitting room, the first time I'd put them on was in my hotel room before my speech, and the ordeal took me a full eight minutes. As it turned out, the size I bought was too small.

I tried to ignore the fact that the band of the top kept rolling down, and as I pulled on my dress, I told myself I could breathe (I couldn't). But I knew how I was really feeling. I was about to stand in front of a crowd of sexy French people—who seemed to never gain an ounce, no matter how many baguettes, cheese plates, or macarons they devoured—and, even though I was a successful author and entrepreneur, deep down I was ashamed.

The Issue with Self-Esteem

Self-esteem is a tricky thing. On the one hand, you can have all the success and confidence in the world, but the shame of being overweight can destroy every inch of your ego. I am a case in point.

Every day growing up my parents told me I was amazing, special, talented, and that I could accomplish anything I wanted. I got a great education, and at school I won competitions for oration, science projects, writing, and business. I was the drum major of the band, helped out at the radio station, and was voted "most revolutionary" my senior year of high school.

After college I had a number of great jobs—and I relished paying my own bills and having money to spare. This small-town girl traveled to L.A., Nashville, New York, Singapore, and Japan. Along the way, I started my own companies, wrote three books, was recognized by *Forbes* as the 5th most influential woman in social media, and worked with Fortune 500 executives. I gave birth to a wonderful child, survived a divorce, and later met the man of my dreams. I am what most people would consider successful, fortunate, and blessed, but there has always been that one thing that prevented others from envying me.

I've been fat and sick most of my adult life.

Many of you are like me. I spent over two decades trying to overcome my struggles with weight and chronic health conditions, and all that time, something in the back of my mind always told me I just needed to accept that I was a loser. The dichotomy of extreme failure and extreme success in my life made no sense to me. If I was as smart and hardworking as anyone I knew, and if I could overcome obstacles and find success in my business and personal life, why was I still always one of the fattest people in the room? I couldn't find answers, and that drove me into self-doubt, binge eating, shame, and self-loathing. It robbed me of some of my joy and forced me to become an expert at compartmentalizing my feelings. Something *had* to be wrong with me, right?

No. Nothing was wrong with me—it was all the dieting advice that was incorrect. Being fat was *not* my fault. As you head into your new life—one that will be changed by fasting—it's time to let go of your shame.

Kicking Shame to the Curb

Letting go of shame isn't an overnight process, and your particular feelings and experiences may run so deep that you'll have to put in significant effort to move past them. But there are a few daily rituals you can perform that will help you make strides toward feeling better about yourself. Try practicing one or all of these every day.

1. Tell yourself the past is the past. Repeat this mantra every time you start to kick yourself for unhealthy habits or behaviors.
2. Write down five good things about yourself. These can be anything, from your beautiful hair to your fierce sense of humor to your compassion and intelligence. Keep this list handy, look at it daily, and keep adding to it.

3. Each morning when you wake up, say one positive thing you plan to do that day. This could be as simple as finishing your to-do list at work or being the best mom you can be. Be vigilant about doing it.

4. Do one good thing for another person or the world every day. Look in on an elderly neighbor, buy the person behind you in line a cup of coffee, or give $10 to your favorite charity.

5. Call one person in your support network every day. You don't have to talk about much; just say hi and that you're thankful for them.

6. When you make a mistake, don't scold yourself. Laugh instead.

7. When you go to bed at night, say, "I am thankful for myself," out loud.

JASON FUNG

Obesity is inexorably intertwined with shame because so many people believe that being fat reflects a person's willpower and character. Obesity is completely different from almost every other disease because there is always the unspoken accusation that you did it to yourself and that you could have done something about it if you weren't such a weak-willed person. Many physicians engage in fat shaming, too, thinking it will give their patients extra motivation to lose weight. I'm always perplexed by this tactic—as if it's not enough that the whole world makes them suffer every single day, the last thing obese people need is to be shamed by a trusted medical professional.

Who deserves the blame for the rise in obesity? I've said it before, and I'll say it again: the "Calories In, Calories Out" model of weight loss. This school of thought may say, "It's all about the calories," but the hidden message is, "Your weight is all your fault."

If you develop breast cancer, though, nobody secretly thinks that you should have done more to prevent it. If you have a stroke, nobody condescendingly tells you to "get with the program." Because of "Calories In, Calories Out" (CICO), obesity has become a disease singularly unique in its association with shame. But the medical establishment, the government, and many diet "experts" are really just trying to deflect the responsibility from the horrible dietary advice they've peddled for decades.

The current mainstream way of thinking—that we limit calories, yet eat more often—has helped cause about 40 percent of the American adult population today to be classified as obese (BMI>30), and 70 percent as overweight or obese (BMI>25). But since CICO is the dominant paradigm among doctors and most of society, the obesity crisis is now believed to be an epidemic of weak willpower. Enter the terrible levels of shame that fat people feel today.

It's time to let this go. Being overweight or obese is not your fault. You've just been getting the wrong information for years. Soon, though, you will embark on a new way of eating, thinking, and living that's going to change your life. I can't *wait* to show you how.

PART III

Your Fasting Plan

CHAPTER 12

Fasting Simplified

EVE MAYER

When I began fasting, I did everything wrong. Sure, the plan worked. But the amount of emotional and physical suffering I endured was unnecessary. I read so much that I was overwhelmed with too many choices, and I wished I had a fairy godmother who could solve all my woes. Of course, that didn't happen, and it took me a year of reading, researching, trying, and failing before I finally figured out what worked best for me.

I'll be honest with you. I like to do the least amount of work for the highest reward. I wish someone had simplified how to fast for me the way I am going to do for you. I'm no fairy godmother, but you can count on me as your blunt best friend, who will tell you the truth about how to ease into fasting and find your groove.

Simple Fasting

Look, I get it. You just want to know if fasting is for you, and you'd like to avoid spending a ton of money, time, pain, and agony to gain the health and body you deserve. I hear you! These wants and our human need to put in the least amount of effort for the greatest reward are exactly why fasting is something you should consider and try.

Why?

Because fasting is simply a manipulation of time. It seems like such a little thing that it couldn't possibly be the answer you have been searching for all these years. But, really, that is all fasting is: how long you need to go between periods of eating to gain the health you desire.

Perhaps you have five pounds to lose or perhaps you have five hundred pounds to lose. I still believe that choosing to control when you eat is a magnificent place to start. So right this minute, adopt a no-cost, no-time-commitment plan to see if fasting might help you at the very lowest level of work. It's okay to be lazy. It's not okay to ignore your own needs, wants, and dreams. You deserve your heart's desire. You hold the key to unlock it.

The most stress-free way to begin fasting is to use a process I call simple fasting—or, because it takes so little effort, lazy fasting. Simple fasting involves starting slowly, with a manageable fasting schedule, then extending it if you like. But if doing minimal work forever is best for you, that's fine, too.

By now, you should have already clearly identified your goal, which is essential in this process. If you haven't, take some time to decide what you are really after in this process. Only after that can you set out on a new lifestyle, with simple fasting guiding you.

So many people feel stressed about choosing the right fasting schedule and making the right decisions about which meals to skip and when to eat. They want to follow strict rules and know exactly what to do. Maybe they've been reading articles and blog posts recommending conflicting fasting schedules, and they feel overwhelmed by all of it. *I've been there.* That's not what I intend to do here. Instead, I want to present an easy entry point with simple fasting.

When you get right down to it, all fasting really means is that you are going to eat less often. Fasting is customizable, and everyone requires a different solution. This is why you have to find what works for you. When I started fasting, I was eating eight times a day, but you may be eating way less often than I was. This is why the series of plans offered in the next chapter is for anyone who wants to find the right fasting lane for them.

I am not a fan of all the words people use to describe various fasts, like 16:8 (meaning you eat only during an eight-hour period and fast during a sixteen-hour span), 36 (a full-day fast), or EF (an extended fast of more than thirty-two hours). I prefer short, plain, easily understandable words because, when you are new to a subject, the lingo that more experienced people use can make you feel like you're hearing a foreign language. So, please, just remember what you read a moment ago. Fasting is simply eating less often. And don't worry. At the end of this book, we do define the many fasting terms you may come across on this journey.

Will I See Unicorns?

Some people experience euphoria and rainbows and peace and visions and freaking unicorns when they fast. If you are one of those people, congratulations! Also, I am jealous of you.

All joking aside, many people do experience increased mental clarity, focus, better sleep, and improved energy within hours of beginning a fast. Others don't—or only feel this after fasting for a few days or weeks. People respond to fasting in all different ways, which is why no one can tell you the ideal schedule. You must try the process and find it for yourself. If you can hit the goals you set for yourself by following one of the easier plans in this book, then you may never need to do a more rigorous fast.

I'm over the suffering and deprivation we've been taught is needed if we want to lose weight. I'm not going to do that anymore. I'm going to eat less often and do the minimal amount of work to reach and maintain my goals. That's right. I'm urging you to be kind to yourself, step into the process, and find your solution by feeling out what's right for you. Imagine that? I now have a healthy body and I'm nice to myself. Honestly, I am my own unicorn!

MEGAN RAMOS

I remember when my husband and I wanted to get in shape, so I found the perfect gym where we could lift weights. When I told my husband, he replied, "Megan, we haven't worked out in two years. I bet we'd both hurt our backs just trying to bend down to touch our toes. Let's instead wake up tomorrow morning and try to stretch."

He was right. The next morning, I started stretching, and I was sore for three days afterward. But by taking baby steps—doing body weight training at home, then months later joining the gym—I was finally ready to lift weights with the big guys.

I recommend approaching fasting in a similar way. It's like developing a muscle, and some people have naturally stronger fasting muscles starting out than others. Some muscles are out of practice, and others bounce back quickly. To work your fasting muscle, you have to have some flexibility, and that's why the plans in this book are laid out from easiest to hardest. Choose what's best for you, but I recommend starting at the beginning rather than diving in with a multiday fast.

Many factors play a role in determining how strong our fasting muscles are. High insulin levels can cause a constant state of hunger that is often difficult to bear, and, other times, medications and health issues (like diabetes) may limit multiday fasts. Our habits play a role, too. Some of us have schedules that require us to eat at certain times of the day (which might limit a daylong fast) and others have beloved traditions they don't want to part with. That's okay. There is a plan for you.

You can strengthen your fasting muscle by keeping two key things in mind. First, build up to your goal gradually with simple fasting. Start slowly, making changes week by week. Taking this gradual approach reduces potential side effects. Second, fast consistently because you're never going to strengthen your fasting muscle if you don't use it on a regular basis. If you go to the gym four times in a year, you can't complain when you don't gain big muscles.

Stress is both good and bad for our bodies. Some stress helps our muscles grow, while too much is harmful. This is the same approach Eve and many others take when it comes to fasting, especially if they're nervous about incorporating it into their lifestyle.

Many people start off with the goal of fasting for thirty-six hours, three times a week. Some do just fine, but they're the exceptions. Most others must start with the basics of just eating three times a day, with no snacking, like our grandparents did. At first, even this might seem like a struggle since we've become such grazers, and we tend to eat more snacks than meals these days. But, eventually, it becomes very easy to stick to your meals and avoid snacking. I had a twenty-six-year-old male client who found all the success he needed by fasting this way. He'd gained thirty pounds in grad school, and while he hadn't developed any diseases, he hated carrying the extra weight. He cut out snacking and started skipping breakfast. Within three months, he lost all his excess weight.

It may feel challenging to cut out one meal, but over time it gets easier. When it's no longer challenging, then it's time to increase your fast just like you would the weight of a kettlebell at the gym. Fasting gets more manageable over time as your body adapts to fueling itself with body fat rather than food. In fact, sometimes you barely notice you're fasting because you don't feel hungry.

Fasting Training Wheels

Most people assume that, during a fast, you should only drink water. That's one way of doing it, but there are many variations that may be just as successful. We encourage people new to fasting to drink bone broth, pickle juice, tea, or coffee in addition to water (still or sparkling). If you love coffee or tea, drink two cups of water for every one cup of coffee or tea in order to stay hydrated. It's also better to drink your tea and coffee black, but even that is a huge change for many, so Dr. Fung and I let our clients add one or two tablespoons of fat to help them adjust (see the full list on page 129).

Salt is crucial to help prevent dehydration because, when you fast, your insulin levels drop. This signals the kidneys to excrete water and electrolytes. Sodium is the gateway electrolyte, so it's critical to replenish it. In short, hydration isn't about just fluid. It requires electrolytes as well. Taking in sodium will cause the body to maintain healthy levels of all other electrolytes.

Think of the products below as your training wheels. It's okay to drink them when you're brand-new to fasting, but you don't want to be using them forever. As the fasting gets easier, you'll find that you need less of these items as time goes on. And you don't have to take them if you don't feel like you need them, either. If you're new to fasting and are nervous, go slowly, challenge yourself a bit each week, be consistent, and use your fasting training wheels until you don't require them anymore.

Your fasting training wheels include:

- Bone broth. Because it's full of vitamins, minerals, and electrolytes, bone broth is great for weight loss, longevity, and disease prevention.
- A homemade low-carb vegetable broth.
- Pickle juice, no sugar.
- Three tablespoons of lemon or lime juice added to water.
- Apple cider vinegar. This is a great appetite suppressant because it mildly decreases your glucose levels. Drink a couple of tablespoons at a time through a straw to protect your tooth enamel.
- Sauerkraut juice.
- Tea and coffee (cold or hot, any variety), with 1 to 2 tablespoons of fat, including heavy cream, half-and-half, whole milk, unsweetened coconut milk, unsweetened almond milk, butter or ghee, or coconut or MCT oil. Avoid sugar and artificial sweeteners.

Common Questions About Fasting

When you start fasting, you're likely to have a lot of questions. We'll discuss many of those issues in Part VI of the book, but we want to address here a few of the most common questions we receive from clients and readers.

Do I have to practice extended fasting for long-term success?
It depends on your fasting goals.

How much weight will I lose when fasting?
Not including water weight, you'll lose about half a pound of body fat per twenty-four hours of fasting. Men and women lose weight at different rates, however. See page 101 for details.

What is the difference between fasting and starving?
Starvation is not voluntary; fasting is. Most people who fast are not malnourished individuals. They tend to be overweight and overnourished.

How long do I have to fast to get into fat-burning mode?
It begins after having fasted for sixteen hours straight. If you fast just by skipping lunch, you're certainly working your fasting muscle, but you're not yet in the fat-burning zone.

What is autophagy and when does it start?
The word *autophagy* derives from the Greek "auto" (self) and "phagein" (to eat). The word literally means "to eat oneself." This is the body's way of getting rid of all the

broken-down, old cell parts that don't serve the body. It's different for everyone, but it starts after twenty-four to thirty-six hours of not eating, increases 300 percent at thirty-six hours, and plateaus at seventy-two hours. The study of autophagy is relatively new and, as such, our knowledge of it is incomplete. Thus, it's recommended if you are fasting for autophagy that you only consume water and salt.

I've heard that fasting can burn muscle. Is this true?

Up to about twenty-four hours or so during a fast, the body mostly uses glucose for energy. During the switch from burning glucose to burning fat, there is a short period when the body uses protein to produce new glucose for energy. This phase is called gluconeogenesis. Many people assume that this period of burning protein is detrimental to health, but it's likely the opposite. Protein is not the same as muscle, so burning protein does not necessarily mean that you are burning muscle. You also have lots of skin and connective tissue. During longer fasts, the body switches to burning mostly fat for energy, and this period of protein metabolism stops.

Can I work out while fasting?

Yes! Movement is encouraged while fasting, and we will discuss this in detail in chapter 15. When you feel lethargic on a fast, the worst thing you can do is do nothing. Walking is great. Vertical movements aid weight loss because they drain the lymphatic system. Some individuals prefer to work out in a fasted state, and professional athletes sometimes train in a fasted state.

I'm afraid that if I fast, I will just pig out later and gain weight. Does that happen?

No. Most people report a dramatic decrease in appetite. Mentally, they may feel hungry, but as soon as they start to eat, they fill up quite quickly.

Will I lose hair from fasting?

Hair loss is one of people's biggest fears because they think that if their hair is falling out, they must be malnourished, but that's not the case. Hair loss is not associated with fasting; it is associated with rapid weight loss. Those who practice intermittent fasting and lose weight steadily—rather than rapidly—don't have issues, as their body is expecting the weight loss. Those who do extended fasts or experience a drastic change in body composition experience hair loss regardless of the type of diet they're on. Hair loss can be stopped by slowing down weight loss, or you can wait it out once the rapid weight loss is over.

Will I have extra skin when I lose weight?

It depends on the individual. Our program has yet to have one client require skin removal surgery since we started in 2012.

How much water should I have daily?

How much water you need varies from person to person, so you should drink water whenever you're thirsty. Thirst is also frequently mistaken for hunger, so drink when you feel hungry, too.

How much salt should I consume daily?

It completely varies per person. Start with $1/4$ teaspoon in your water and continue to add more until you stop experiencing headaches or lethargy.

Can I drink alcohol while fasting?

We don't suggest it. Alcohol puts you at risk for dehydration and can raise your insulin levels, which you're trying to lower during a fast. We even recommend abstaining from alcohol—except, if you wish, a glass of wine—during your first meal after a fast. If you do drink, I recommend dry wines (red or white) and spirits as long as the mixers don't have sugars or sweeteners in them. For example, a vodka soda with lime won't raise your insulin like, say, a margarita. Also, we encourage people to limit themselves to one drink per day. If you drink more than one alcoholic beverage, alternate the booze with a glass of water. Finally, it's better to consume alcohol during rather than outside your eating windows (say, during the time you've set aside for dinner).

Can I have sugar or stevia while fasting?

This is a hard no. Both sugar and stevia will stimulate the production of insulin, which you are trying to reduce with fasting.

Can I take my vitamin or daily supplements while fasting?

Supplements may hinder autophagy. When fasting for metabolic reasons (meaning insulin resistance–related conditions such as type 2 diabetes, obesity, PCOS, and nonalcoholic fatty liver disease), the effectiveness of supplements is questionable. Most vitamins are fat soluble, but if you're not taking in fat they won't be as effective. Probiotics are fine to continue taking while fasting.

What do I do about medications I normally take with food?
Talk to your doctor.

CHAPTER 13

Stop Snacking

EVE MAYER

Your first major step toward fasting is to stop snacking—for good. If you're like I was, this may be the most intimidating part of fasting. Trust me, though. If I could do it, you can, too!

When I used to be hungry, I snacked. I snacked *a lot*. How many times a day on average do you think you eat? Three or four times? You might be surprised at the truth. Seriously, stop for a minute and add up how many times you ate yesterday and try to think back to the day before as well.

Yes, the two pieces of candy from the candy bowl count as one time. Yes, the biscuit you had with your coffee counts. Yes, the half chocolate bar you ate when you needed a little something sweet late at night counts. By the time you add all of this up, I bet you reach at least six or seven teeny-weeny, innocent, mindless snacks.

Many of us have been told that we need to eat small meals and snacks throughout the day to keep our metabolic engines running. Well, guess what? I snacked and snacked like a champ. I did my very best to keep that engine running while eating smaller portions, and, shockingly, my hope that these tiny meals would keep my metabolic engines burning didn't work for me. I got fat. I tried diets, and I stayed fat. I had bariatric surgeries and I remained (a bit less) fat. My bariatric surgeries made my stomach smaller, so I ate smaller meals more often. Yet I never, ever felt full for much more than an hour. The never-ending cycle of hunger, food selection, food preparation, eating, cleaning up, and repeating took up a huge portion of my life.

When I learned about fasting, I thought it was a ridiculous idea. Determined to prove Dr. Fung's theories wrong, I jumped in headfirst with a thirty-six-hour fast. Obviously, over time I realized Dr. Fung was right, and fasting helped me to discover a new, wonderful, healthy life. But had I been smarter, I would have progressed at a more reasonable rate doing what Megan suggests: treating fasting like weight lifting, gradually building my fasting muscle.

It's easy to start. Are you ready? Stop snacking. Right now.

Step One: How Not to Snack

The rules are simple. Eat three meals a day until you are full. Each meal should take no longer than one hour, and you get to choose the time of the meals. You get to pick the food in the meals. If you drink soda or any kind of drink, even diet drinks, start limiting them to your mealtimes. If you must have gum, have it after you eat and within one hour of a meal.

But, *no snacks*, which means: no gum, candy, mints, sugar, food, juice, sweet drinks—natural or otherwise—broth, smoothies, or sports drinks, except at mealtimes.

If this sounds easy to you, then I completely admire you. And, pardon the pun, but I bet fasting might just end up being a piece of cake for you. If you're a bit older, like myself, not snacking may even sound familiar. I remember being told that if I had a snack after school it would spoil my dinner. Eventually, the prevailing wisdom changed, and parents—mine included—gave their children snacks after school.

Baby Steps May Be Necessary

Some of you reading this may realize that you eat eight to ten times a day. If that's the case, don't stop snacking cold turkey. If you do, you'd be eating only a third of the times that you currently do. This compares to a person smoking a pack of cigarettes a day going to just a few cigarettes a day. You're likely to fail quickly, and I don't want that to happen.

Instead, cut out one snack time per day for a week, just like this:

- **WEEK 1**: Go from eating eight times a day to eating seven times a day.
- **WEEK 2**: Go from eating seven times a day to eating six times a day.
- **WEEK 3**: Go from eating six times a day to eating five times a day.

Get the picture?

When I stopped snacking, it was uncomfortable, but it was not painful or unbearable. It just felt foreign because I was used to eating at certain times of the day, and my body and mind knew it. At first, I felt hungry during the periods I would usually eat. But my body began to adjust and didn't expect the food I had been giving it at that time slot. When I skipped my usual snack, I also felt hungrier than normal at my next eating period, and I would eat a bit extra then. But my body adjusted to this over time, too.

I now avoid snacking about 90 percent of the time, and it's one of the reasons I feel that I've been able to maintain a healthier weight for the first time in my life. Sometimes, I can't resist sneaking in something extra delicious, or if I have a hunger that lasts for hours, I may have a snack. But this is rare. Thankfully, now that I'm used to it, staying away from snacks isn't as hard as I thought it would be. I understand that fit people do the right things *most* of the time, and that is what I strive for.

JASON FUNG

Controlling hunger is crucial to weight loss, so how do you rein it in? The standard dietary suggestion given to most people by doctors and diet books is to eat six or seven small meals per day because, if you can prevent hunger, then you may also be able to make better food choices and avoid overeating. On the surface, this advice seems pretty reasonable. But dig a little deeper and it falls apart.

The most important determinant of how much you eat is how hungry you are. Yes, you can deliberately eat less, but you can't *decide* to be less hungry. If you are constantly eating, but are still hungry, it takes a toll on your health, day after day, month after month, year after year. You are in a never-ending battle with your own body. However, if you are less hungry, you'll eat less. You'll be working with your body to lose weight, not fighting against it.

Is there any scientific evidence to suggest that snacking or eating small meals constantly will prevent hunger? That would be a big NO. Let me repeat that. Science does not support the idea that eating nonstop will alleviate the gnawing, aching grumbling in your stomach that leads you to snack. It's also not necessary to snack because our body stores food energy (calories) as body fat for the exact reason of providing calories when we need them. Finally, eating all the time is sort of a nuisance. If you are trying to find food six or seven times per day, when are you supposed to get anything done? You are constantly thinking about what you need to eat and when to eat it.

You might believe that grazing or snacking will prevent overeating. If this were true, what is the point of an appetizer? An hors d'oeuvre is served to make us eat *more*. Eating a small, tasty morsel makes us hungrier because it causes us to start salivating and thinking about food. In French, this is called an *amuse-bouche*, meaning "something that amuses the mouth." Why? So that we will eat more. Giving too large a portion would stimulate the satiety hormones and dull the appetite. But a tiny portion does the opposite.

Now, think about a time where you weren't really hungry, but it was breakfast time, and you ate because people have always said breakfast is the most important meal of the day. To your surprise, after you started eating, you finished an entire meal. Before you first lifted your fork, though, you could have easily skipped the meal and been fine. Has this happened to you? It's happened to me many, many times, and I've noticed it simply because I've conditioned myself to think about hunger. In short, eating when you are not hungry is not a good strategy for weight loss. Yet people are continually scolded for having the temerity to skip a single meal or snack. This flies in the face of all logic.

If we eat small meals six or seven times per day, we whet our appetite, then deliberately stop before we're satiated. We then repeat this multiple times per day. This is not going to decrease our appetite, but it will increase it—a lot. And because we're hungry but haven't eaten our fill, we must exert a significant amount of willpower to stop ourselves from snacking. This is exhausting, yet, day after day, it continues.

The way to break this cycle—and continuously give yourself a smaller appetite—is to eat less frequently and stop snacking.

CHAPTER 14

Stepping into Fasting

EVE MAYER

I know you're ready to move forward with your life in a transformative way. You've been able to stop snacking (step one), so it's time to head right into fasting. Don't worry, it's not that difficult! We'll ease you in, and as we said before, there is no need to move on to a harder next step if you're not ready or able. Stick with the step you're most comfortable with and continue on when you've mastered it.

Step Two: Skip Breakfast

Yep, that's it. Instead of eating three meals a day, you will eat two. And remember, no snacks.

How can you prepare the night before you skip breakfast? Great question. Your dinner should consist of healthy, whole foods that will fill you up and make you feel good. Preferably, stick to a low-carb diet, like we described in chapter 5. When you eat sugary foods or other products you've decided are not great for you, they can make you feel hungrier the next morning. I find it helpful to consume lots of healthy fat the night before fasting, like meat or cheese, plus a big green salad loaded with vegetables. You should eat this meal within one hour, a time period that allows you to enjoy what you're eating and who you're eating with—and get full.

When you skip breakfast the next day, you will be hungry at first because your body is used to eating in the morning. Distract yourself by hydrating instead. Drink water, carbonated water, tea, or coffee. If you must, have a splash of heavy cream in your tea or coffee. I skip breakfast every morning, but I drink two cups of coffee, and they set me right. Do not use sugar or sweetener of any kind, including natural sweeteners like stevia. Why? Because for many people that sweetness can very well trigger hunger and make fasting more difficult than it needs to be. Remember, weight loss is about controlling hunger, not calories. When it's time for lunch, eat within your one-hour eating period, then wait no more than eight hours until dinner, when you should dine within one hour again.

There was a time when I woke up and ate as soon as possible, hoping to get my metabolism running. Now, I rarely eat in the morning, and it has come to be no big deal. In fact, I was shocked to find that once I was used to it—which took me about six months—I hardly ever felt hungry. If I did, my trusty coffee with a splash of heavy cream did the trick.

How often you choose to skip breakfast is completely up to you. If the first day feels like no big deal, skip breakfast again the next day. If not eating is hell, only do it once or twice a week until it is bearable. Eventually step up your frequency (I recommend adding one day a week, day by day) until you reach a place that's comfortable to you.

You may have an unusual schedule and want to know if you can skip dinner instead of breakfast. Of course you can! The meal you choose not to eat doesn't matter. The goal here is for you to get comfortable with eating only two meals a day, each consumed within one hour, separated by no more than eight hours between the two meals.

Can you still work out on the morning that you fast? Yes, you certainly can. I do, and I find I exercise so much better while I'm fasting than I do when I work out after eating. Another unexpected surprise for me about skipping breakfast is that now my family has more time to sit together on the sofa, sip coffee, and give the doggy extra belly rubs. Plus, we don't have to do breakfast dishes before we run out the door, and we save money on groceries. Win-win!

In short, skip breakfast as frequently as you wish and see what changes happen in your body and mind. I know many people who do this daily, and this is the total extent of their fasting practice. I've seen these people lose weight, get off medications, and change their lives for the better. This may be the case for you, or you may find after a time that you wish to go further. Again, it's up to you. This is *your* life.

Step Three: Skip Lunch

Now that you are never snacking and regularly and comfortably skipping breakfast as often as you wish, you're ready for a day where you only eat an evening meal.

If you think about it, skipping lunch after not eating breakfast only adds about six more hours of fasting. For most people, it also isn't nearly as traumatizing as they think it will be. Plus, you only have to *try* for *one day*. The following day, you can go back to the eating schedule you normally follow.

Yes, you are going to get hungry at lunchtime, but keep yourself busy, hydrated, distracted, and determined to reach your goals. When dinnertime rolls around, eat healthy, filling foods for a period of one hour, and make sure to finish up at least two hours before you go to bed.

If eating only dinner for a day isn't that difficult, try it again the following week. If the process continues to be easy for you, begin eating only dinner two days per week but not on consecutive days. After a period of time you feel comfortable with, decide if you want to eat only dinner three days a week, nonconsecutively.

Does it matter which meals you skip? Just as with step one, no. You can eat at the mealtime of your choosing. I have suggested eating only at dinner because this is the meal most commonly shared by family or socially. If you have an untraditional schedule, feel free to go with just breakfast, or you could choose to eat only lunch. The only necessity is that you eat a meal with a fasting period of twenty-three hours before it. So, if you decide to eat only lunch, your last meal the day before would need to be lunch.

As a person who had zero experience fasting before I dove into this new lifestyle, I never found a twenty-four-hour fast too mentally or physically difficult. I'd worked up to it by skipping snacks and breakfast, so one more meal didn't seem terrible. I still feel this way, and now I skip breakfast most days.

Tracking Your Fasting

Over the years I've seen many people track their twenty-four-hour fasts using a timer or an app. They find it's not only helpful but gives them a feeling of accomplishment to see the hours they've fasted climbing—from fourteen to eighteen to . . . time's up! Dinner's ready!

If this helps you, go for it! Personally, I found tracking my fasting hours not to be helpful. I began to focus on how long I had left and then dwelled on the fact that my belly was empty. It caused me to doubt if I could finish the fast, and I realized that the busier and more distracted I became, the better I could enjoy the process. But you may be different!

Step Four: Skip Dinner

I'm going to be honest; I think that step four is the most difficult. But if you have stopped snacking, often skip breakfast, and are at ease with bypassing lunch a few times a week—yet still aren't hitting your weight loss and health goals—then it is time for a thirty-six-hour fast.

I know, I know! You may be thinking, *Wait. Thirty-six hours? I thought I was only fasting for a full day?* Stop for a second and think: This fast is referred to as a 36 because you will not be eating for thirty-six hours. You finish dinner at, say, 7:00 p.m. You sleep for the night. You don't eat the following day. You sleep again. You wake up and eat breakfast at 7 a.m. Boom, you just fasted for thirty-six hours!

I didn't use the schedule above for my first thirty-six-hour fast, and I found it extremely difficult. Why? Because I was awake for a lot more of it, obsessing over my hunger. I ate breakfast and then began a thirty-six-hour fast, which means I was only asleep for eight hours of it. It is *so* much easier to do a thirty-six-hour fast when you are asleep for sixteen hours of fasting time. This means you only deal with hunger for twenty hours, or the hours that you are awake.

I recommend scheduling this fast on a day you are as busy as possible and around as little food as possible. Set yourself up for success, not suffering, and do your best to limit mealtimes with others, shopping, and cooking. Because these things are sometimes unavoidable, you should always ask for support, and you may just be surprised at how helpful people will be. I also tend to sleep less when I fast for a full day, and while this was once confusing and discouraging, I now view it as an opportunity, and I plan to get extra things done when I'm not sleeping.

Finally, remember that not eating for one day is called *fasting*, not *starving*. Choice is the differentiator, and while going to bed with no food in your stomach may be scary, once you're asleep you won't even notice. Plus, during the day you can practice all the tricks you did before: hydrate, find an activity you enjoy, and remind yourself that you're doing awesome things to reach your goal.

Will you be hungry? Yes! You aren't used to going a day without food, and your body is going to get hungry in response to the behaviors it has been used to. But remember that hunger isn't a bad thing; it's your body telling you it's burning fat. Plus, you're heading into uncharted territory here, and every time you fast is a singular experience. When you lengthen a fast for a longer period than you have ever done before, you will probably feel hungrier. But you can handle it because you've built up to it.

The frequency of a thirty-six-hour fast is completely up to you. If you find your first 36 traumatic, wait one month before trying it again. If you think it's pretty darn easy, do it again the following week. As the thirty-six-hour fast becomes more seamless, you can use it as a tool to speed up your progress toward your goals.

Tips for Getting to Sleep After a Full Day of Fasting

1. Remind yourself that the sooner you get to sleep, the sooner you will eat in the morning.
2. Make sure you are well hydrated. Some people find a mug of warm caffeine-free tea (like chamomile) helps them to feel fuller.
3. Consider melatonin supplements if needed.
4. Warm baths are great but are most helpful at least a few hours before bedtime.
5. Avoid computer, TV, and phone screens for a couple of hours before bed if possible.
6. Try meditating, deep breathing, aromatherapy, reading, or whatever makes you feel restful.
7. Decide what you will eat for breakfast. Don't prepare it because you might end up shoving some of it into your mouth accidentally that night (it has happened to me!). Knowing what you are looking forward to enjoying foodwise in the morning can make the hunger more bearable.

Step Five: Extend the Fast to Day Two

Just look at you! You have become an impressive fasting beast. You stopped snacking, often skip breakfast, can also skip lunch when you choose, and can even go a full day with no food. Guess what? You've already completed an extended fast because most people define one as being longer than twenty-four hours. Nice going!

So how does one prepare to not eat for *more* than one day? First, you must decide if you want to do it at all. So many people have lost weight and met all their health goals without ever doing an extended fast, and that is more than fine. For example, one woman I followed online stopped snacking and skipped breakfast every morning. With this schedule, she lost sixty pounds in a year, and her doctor took her off her medication for type 2 diabetes because her blood sugars were in a healthy range. Another man I followed lost over one hundred pounds over a nine-month period by skipping breakfast every day and forgoing lunch on two of those days. He also lowered his blood pressure and was taken off the medication he had used for eighteen years.

This is not the case for everyone, though, and there are many reasons you might want to try an extended fast. They include:

- You want to speed up the process toward your goal.
- Your weight and measurements have stalled for a month or more.
- You are curious to see what it is like.
- You are seeking added health benefits beyond just weight loss. For example, extended fasts are more effective at lower insulin levels in those with type 2 diabetics. After thirty-six hours, ketosis begins, and after forty-eight hours, autophagy, or the process of cellular cleansing, kick-starts. Extended fasts are also reported to heighten mental clarity.

To strengthen your fasting muscles and see if an extended fast will work for you, start by not eating for a whole day, sleeping, and then fasting until lunch the following day. If this feels comfortable, a few days or so later, try forty-two hours. Then forty-eight. Later, step up to seventy-two hours, then five days. The important thing is to increase your fasting window bit by bit. Once one fasting regimen becomes easy, then extend it to the next level of fasting.

It is tempting to think that the longer the fast, the better. This simply isn't the case. I have done two extremely long fasts: one for eleven days and one for ten days. My results were impressive: I lost about twelve pounds each time, and my skin glowed like I'd had a $300 facial at a fancy spa. But the fasting was very tough mentally, and my weight loss was the same as it would have been if I'd just fasted for ten single, nonconsecutive days over a month. Other than those, I have done about twenty extended fasts, most being thirty-six to forty-eight hours. They can be tough for me to stick to, but if I schedule them when I am busy and when my husband can cook and grocery shop, I can usually complete them.

On my eleven-day fast, I was shocked at all the thoughts and emotions that went through my head. I kept a diary, and I sometimes go back and marvel at how psychologically amped up I felt. My most powerful emotion was anger. I was mad at my trainer—whom I'd known for two years—telling me she didn't feel comfortable with me working out midway through my fast. I was angry at the fact that I wore my husband out by telling him almost hourly about all the new data I'd found about fasting's success rates. I was upset that, some nights, I didn't sleep well—or woke up at 4:00 a.m.—because I had so much extra energy. I was angry when I realized I was going to have to start living a new life that wasn't focused on food, and that would require me to develop new skills and interests. And, most of all, I was angry that I hadn't tried fasting earlier. I'd spent *years* struggling and failing to lose weight and failing to feel better, and, yet, during an extended fast, I felt healthier and more energetic than I had in years. It was like I was living a new life! I kept thinking, *Why didn't anyone tell me about fasting earlier?*

When you decide to do an extended fast—especially one that's more than forty-eight hours—I think it makes sense to consider having a doctor involved. When I did my eleven-day fast, I was under the care of Dr. Fung and Megan, and I also informed my primary care doctor, who was supportive of my exploring new methods for my health. As I said in chapter 10 as well, a doctor's care becomes especially important when you have health issues or medications you normally take with food.

Regardless of what you decide to do, extended fasts can be a wonderful reboot for your weight and overall health. As part of a larger fasting plan, they can be life-changing for any number of health conditions, and they may even change your perspective on your health, happiness, and self-worth.

JASON FUNG

Now that you know the many ways you can begin a fast and proceed through it, it's time to learn what to do when you break your fast.

For most fasts—that is, five days or less—you don't need to worry too much. Just eat slowly and mindfully. As you know by now, once you start eating, your body will tell you to keep eating until your hormones signal satiety. Don't eat beyond that point. Your meal should be made up of colorful, nutritious, healthy foods that make you feel good. Chew slowly and deliberately, and do not stuff food into your mouth. The first bite of food may feel strange—or, conversely, taste more delicious than anything you've experienced before. Just sit with that feeling and don't overthink it. Plan to eat your meal within one hour, and remember to drink plenty of liquids over the course of it.

After longer fasts of five days or more, I often advise people to plan a small snack—like a handful of nuts or a small salad—about thirty minutes ahead of the main meal. The small snack is a form of refeeding. When doing longer fasts, there is a small risk of something called refeeding syndrome, where a large meal upon breaking a fast causes electrolytes to move into the cells too rapidly. This can be very dangerous, and refeeding slowly reduces that risk.

The second reason for refeeding is to prevent the overeating that some people do when they break a fast. It gives the digestive system a chance to "warm up," if you will, since it hasn't been used for many days. For shorter fasts of fewer than thirty-six hours, it makes little difference.

CHAPTER 15

Exercise for Health, Not Weight Loss

EVE MAYER

Note: If you love your workouts and are more or less at the body weight you want, you can probably skip this chapter and go straight to chapter 16.

However, if you groan at the thought of working out, are struggling to fit exercise into your schedule, or have never achieved the fitness results you hoped for, then keep reading!

My Struggles with Exercise

Many years ago, when I weighed about 280 pounds, I decided to train for a three-day, sixty-mile walk in support of the fight against breast cancer. Besides raising money for this worthy cause, my preparation called for me to walk three to fifteen miles per day, several days a week, for six months.

Have you ever exercised carrying around an extra 30, 60, or 100 pounds? I have, and then some! At 280 pounds, I was dragging around an extra 130 pounds, so you might as well imagine me carrying an adult woman on my back. One mile of walking was certainly uncomfortable, but months of walking at that weight were agony. Everything chafed and hurt, and my feet strained to support the weight of my body. I spent every weekend walking up to ten miles, while joggers and speed walkers zipped by me. I felt sluggish, slow, and ashamed. Yet I persisted, trained, and was determined to complete the walk on behalf of the women I knew who had suffered through breast cancer so bravely.

Although I trained at home in Dallas on mostly flat terrain, the walk I'd signed up for with a group of old college pals was from San Jose to San Francisco. Let's just say California is a slight bit hillier than my training had prepared me for! The locals said they were small hills, but they immediately felt more like giant mountains to me.

For three days, I lined up near the front of the group with thousands of other participants behind me. Yet without fail, by the end of the day I was always one of the last people on the course, if not the very last. There was a sweeper van at the end of the crowd of walkers that picked up participants who needed to be transported to the next water stop. There was also a kind volunteer who rode her bike behind the very last walker. The biker and I became very well acquainted, and by day two, she "kindly" suggested that perhaps I should take advantage of the van just to get a little break. I "politely" declined, and as morning turned into afternoon, she practically forced me into the van.

The three days of the walk were humbling at best and mortifying at worst. People of every age and size passed me, from fit walkers to not-so-fit walkers to a few ladies who were even larger than I was. One day I looked up to see a woman who was likely in her seventies, her bald head suggesting she had cancer, breeze past me, sure-footed in her stride. I'm not sure if I imagined this, but I swear I also remember an elderly woman on crutches cruising by, eyeing me with pity.

I'm sharing this story to show you that, all throughout my obesity, I exercised. For years, I spent money, time, and hope on weight training, dance classes, walking, yoga, gyms, personal trainers, Pilates, and more. Yet exercise was never an effective way to help me lose weight. In fact, during my six months of training for the three-day walk, I gained five pounds. Perhaps it was muscle, you ask? I'm not so sure. My clothes were tighter, so I think it was fat.

In order to lose weight and regain your health, you need to add something else to your lifestyle other than exercise. That something is fasting.

Exercising While Fasting

Many people are shocked to find out that working out while fasting is not only allowed but encouraged. Since I started fasting two years ago, I'm now in the best shape of my adult life, with more muscle than ever. I continue to lose fat and gain muscle at a slow yet steady rate, and I'm filled with vitality almost every day.

I'll tell you a secret. I spend less time exercising than I used to. In the past, I used a walking desk at least three hours a day, plus I worked out with a trainer a few times a week. I was in worse shape, and I was fatter and much hungrier. Today, I actually enjoy fitness—most of the time!

My routine is simple. Most days I walk the dog at a brisk pace that gets my heart going for about twenty minutes, and I do weight lifting with barbells or machines about twice a week for forty-five minutes each time. When I say weight lifting, I want you to picture very basic, small weights or machines set at low resistance, under the supervision of a personal trainer or using a workout app with my husband at the gym. Recently, I added CrossFit into the mix two times per week. I'm often the worst in the class, but I feel like a stud doing squats and lifts with barbells, even if the weight I'm using isn't substantial.

In short, with fasting, I'm exercising less, but I am more fit than ever before.

Should you exercise? Yes! Fitness benefits your brain, your breathing, your heart, your lungs, your muscles, your digestion, your cellular composition, your mental health, and more. When I abstain from exercise completely, I find that my mood and contentment level start to dip after about ten days. Once again, finding the right exercise for you is a personal journey, and while Megan will lay out some recommendations, no one should tell you with assuredness that three hours of any specific activity a week is exactly what you need.

Recently, my husband and I went on vacation to Nevada. We traveled to a tiny town called Blue Diamond, which is nestled in the desert mountains and has a charming restaurant, a bike shop, and 150 miles of rocky trails meant for biking, horseback riding, or walking. My husband and I made the very uncharacteristic decision to head out on a hike—the first real hike of my life. I was nervous, to say the least! But I forced myself to think positive and do it. The two-mile trail took one and a half hours, and when we reached the top of a peak, we were rewarded with stunning views of red- and brown-striped mountains and a crisp, cool wind on our faces.

Suddenly, I was overcome with emotions, and I began to cry.

I'm sure you've seen pictures of friends at the top of some beautiful mountain, their arms reaching into the air in celebration. Many of my friends have posted these photos online, and I've always felt happy for them that they were strong enough to reach the peak. But I knew it was something *I'd* never be able to do.

When I climbed to the top of that Nevada mountain, suddenly and unexpectedly, I proved myself wrong. You can do the same!

1. **CHOOSE SOMETHING YOU CAN AFFORD AND THAT FITS IN YOUR SCHEDULE.** Don't sign up for a marathon if you can only devote two hours a week to running. Don't enroll in expensive private Pilates classes that your budget will not allow.

2. **FIND SOMETHING THAT YOU ENJOY.** If you are groaning now, I get it. I used to hate any and all kinds of exercise because getting up and moving was painful. I chose walking because I enjoyed the peace of being outside, and I worked my way into it slowly. If you can't find an exercise you like right away, then choose to exercise in an environment you enjoy. For example, if you're looking for a low-impact way to reduce stress, try gentle yoga. If you love the pool, sign up for a water aerobics class.

3. **CONSIDER DOING LESS CARDIO AND MORE STRENGTH TRAINING.** Many people who haven't found a great exercise routine find cardio intimidating because it wears them out quickly. You can work your way into strength training slowly and build up. Seeing progress as you add more and more weights to your routine is *so* satisfying.

4. **ONCE YOUR EXERCISE ROUTINE BECOMES EASY OR BORING, ADD SOMETHING TO IT, MAKE IT MORE DIFFICULT, OR CHANGE THE ENVIRONMENT.** Your body is highly adaptable, and without constant change, your workouts will decrease in effectiveness.

5. **EXAMINE YOUR MOOD EACH WEEK AND LOOK FOR A CONNECTION BETWEEN YOUR WORKOUTS AND YOUR MENTAL STATE.** Do you feel calmer when you take walks outside? Do you feel happier when you lift weights a few times a week? Many people find that exercise is a huge influence on their emotional state.

6. **GET AN EXERCISE BUDDY.** Exercise is a great excuse to spend time productively with someone you enjoy. This could be your spouse, partner, friend, boss, mom, or neighbor. It really doesn't matter. If you are both fasting, this will also help to replace some of the bonding time you may have spent together in the past eating.

7. **OPTIMIZE YOUR WORKOUTS.** Compare your workouts when fasting and feasting to find the ideal time to do them. I personally have a much better workout when I fast, but some people may see better results when feasting.

MEGAN RAMOS

I used to hate everything about exercise: the sweating, the exertion, the "burn . . ." *Everything.* Now that I'm healthy and energetic, I don't mind it. Some days, I even feel like I fit in at the gym. I've focused on following my trainer's program, and I've made great progress each week. I've gained muscle, and I've added bone mass, which has been especially important because I was diagnosed with osteoporosis at age thirty.

Exercise has also been a great outlet for my stress. I used to cope with stress by eating my weight in pizza or pasta, which left me feeling emotionally and physically awful. Today, I can be having a terrible day, and I know that I will leave the gym feeling renewed. Many times, I've walked into the gym in tears on a rough day, and left practically dancing down the street, ready to conquer the world.

But, like Eve, it took me a long time to get there. For nearly a decade I spent thousands of dollars going to the gym with the sole intention of losing weight, and I always failed miserably. Today, I work out to relieve stress and to get stronger, but not to lose weight.

Let me say that again: Do not go to the gym to try to burn calories. Do not get on the stationary bike like I used to, turn on the calorie counter, and step off when you've cycled off the 250 calories that were in the glazed doughnut you had for breakfast. Weight loss is *not* about "Calories In, Calories Out," so while exercise will deplete calories, it will not reduce fat. That's why I never lost weight. I wasn't broken, I was just misinformed.

I'm not the only person who's been in the dark about this. Many of the people I meet tell me that they're exercising harder than they have in their lives, but they're still gaining weight. They may be consuming only 800 calories a day and burning 1,500 calories a day, but the pounds pile on. What is going on here? The issue is that the "Calories In, Calories Out" model of obesity is extremely flawed. As we've discussed, weight changes are hormonally regulated, with insulin playing the key role in fat storage.

The truth about exercise and weight loss didn't fully hit me until I went to Paris shortly after I lost 60 pounds by fasting and following a low-carb diet. I spent several days roaming the streets with my friends and didn't see a single gym. Not one. Yet the locals weren't obese. In fact, you can't even be obese and function normally in Paris. There are no escalators. The elevators are *tiny*. I am 5 feet 3 inches tall and was 120 pounds at the time, and I got claustrophobic in one. The seats on the subway are narrow, and so are the hallways.

I was traveling with a friend who was quite overweight, and she said to me, "Megan, I'm never coming back here until I've lost weight. I didn't realize how obese I was until this trip. I don't get it. The locals don't have breakfast. They eat twice a day, and their diets are full of fatty foods like cheese, butter, and eggs. Parisians walk a lot, but they don't work out otherwise. And they're so skinny!"

Exercise won't help you lose weight. Obesity is hormonally regulated, so it doesn't matter how many minutes you spend on the treadmill. If you're eating the wrong foods too often and not giving your body a chance to burn fat, you're not going to lose any weight. Yes, you can exercise your muscles. But you can't exercise your hormones.

Most of my clients have been completely brainwashed into thinking that they won't lose weight unless they work out, or that exercise will somehow speed up the process. This notion has failed repeatedly, but it's something that is so difficult for people to let go of. Yes, exercise has many other health benefits, including improved muscle tone, mood, flexibility, and stamina. But it will not improve your weight loss.

For those other health reasons, I drag myself out of bed four times a week at 4:30 a.m. to lift weights. I walk my two greyhounds for at least an hour every day. I've happily gained several pounds of lean mass by doing so, and I now have strong, healthy bones. But exercise didn't help me lose weight before, and it doesn't really help much now.

The "Calories In, Calories Out" model of obesity is completely broken, giving people the illusion of control. I'll often ask clients why they think working out will help them lose weight when it hasn't before. They always answer that they haven't been working out hard enough.

Losing pounds *is* in your control, but it has nothing to do with an aerobics class. It has everything to do with what you eat and, especially, when you eat. Although I encourage all my clients to be as active as they can, we should not be exercising to lose weight. We've tried it over and over, and we've just gotten fatter. So, please, exercise *and* fast.

JASON FUNG

One of my most famous clients is UFC champion Georges St-Pierre. On November 4, 2017, this mixed martial arts legend was set to fight Michael Bisping in the UFC Middleweight Championship. Georges hadn't fought a title bout in four years, so the stakes were high—for him, his fans, and the sport. He needed to put on pounds before the weigh-in, so he began eating every two hours. Throughout his life, he'd been conditioned to think that, to build muscle, you need to consume greater quantities more often, so, in addition to his rigorous workouts, his food intake was at an all-time high. The results were disastrous. He began having cramps, he couldn't sleep, he regularly threw up his breakfast, and, most alarmingly, there was blood in his stool. Unable to find a diagnosis, Georges literally muscled his way through the pain. Against all odds, he managed to win the championship in the third round.

Afterward, doctors finally determined that Georges was suffering from ulcerative colitis, an inflammatory bowel disease that causes ulcers in the intestines. He was prescribed medication, but, in addition to it, Georges wanted to find another way to deal with his symptoms. He came to me, and I recommended intermittent fasting. Within weeks, Georges's symptoms vastly improved. He began sleeping better, his inflammation subsided, and he stopped experiencing cramps. His bone and muscle mass improved, and his fat percentage went down. Best of all for an elite athlete like Georges, his energy was up, and he felt his workouts were better. Today, Georges starts his day by training on an empty stomach, and he says he feels lighter on his feet, with better concentration and more focus.

The truth is that exercising on an empty stomach will give you a better workout than exercising after you've eaten. Think about it: Would you rather wrestle with a hungry lion or a lion who's just stuffed himself with a big, juicy antelope? You'd never choose the hungry lion because he'd be fierce, fast, and out for the kill!

When you train while fasting, your insulin levels are down, but your noradrenaline and growth hormone levels are up. In this hormonal state, you have more energy (because noradrenaline and HGH provide it) so you can work out harder. And when you do eat after working out, your HGH stays high. HGH helps rebuild muscle, so you recover faster.

There is no physiological reason a person needs to eat before working out. In fact, I often encourage people to work out first thing in the morning on an empty stomach because they already have all the fuel their body needs. At 4:00 a.m., your body experiences a surge of hormones that allows glucose to enter your bloodstream. When you wake up to work out at 6:00 or 7:00 a.m.—or any time in the morning—your body is still burning that glucose. You don't *need* to eat.

Working out in conjunction to fasting may lead to a leaner, more muscular, more energized you. Just remember, it's not the exercise alone that's helping you lose weight. It's the fasting.

CHAPTER 16

Feasting Without Guilt

EVE MAYER

The act of fasting is choosing to refrain from eating for a specific amount of time. Feasting should be the act of choosing to eat for a specific amount of time. Forget the extended, gluttonous hamburger-, pizza-, soda-, French fry–, and ice cream–soaked feasts you've indulged in in the past, though. I'm going to show you how to feast mindfully, healthily, and satisfyingly throughout your new, fasting-inspired life. Because, yes, feasting *is* allowed when you're not fasting—though it will feel a lot different than it did before.

What Does a Feast Look and Feel Like?

When I was first learning to fast, I was terrified of the unknown. Would I feel hungry? Would I starve? Would I even be able to *do* it? After I'd completed a few short fasts, though, I started to know what to expect. I became comfortable with a little bit of hunger, and I stopped being startled when the hunger ceased even though I hadn't eaten. There are times when I am fasting that I feel incredible, and fasting has helped me have moments when I've experienced clarity of thought and joy at a higher level than I did in the past.

I still look forward to eating, though, and I still enjoy eating *a lot*. I adore everything about feasting, from the smell of the food, to the texture and taste of it in my mouth. Before I started fasting, I assumed that people who skipped meals just didn't like eating that much, and that they were weirdos who could never be foodies. I worried that, if I fasted, I would become someone like that, too.

I was so wrong!

For every fast, there should also be a feast. Let me be clear that feasting is not what I used to believe: an extended period of gluttonous eating until I was in pain from my stomach begging for mercy. Instead, a feast should be a period of time—around one hour, ideally—when you eat healthy foods joyfully and mindfully until you are full.

I am not suggesting that if you abhor broccoli you should control your emotions enough to sit down to a big bowl of broccoli with a lustfully ravenous look in your eye. Ideally, you should pick low-carb, healthy-fat foods that you *enjoy* eating, preferably whole and organic. It is true that you won't love everything that is healthy for you, but over time, your preferences and tastes will change.

And this is a crucial change.

Feasting does not mean eating perfectly, though. Even fit people sometimes indulge in unhealthy foods. Today, 90 percent of the time, I eat what I know makes me feel great, and 10 percent of the time, I have something that has no nutritional value and is not likely to make me feel good. I freaking love hamburgers, and I wish I could eat them as often as I used to. But I know that if I eat fries, a double-bacon cheeseburger, and a chocolate milk shake, I will feel sluggish and gain weight. I also notice that when I eat this way, I have incredible, undeniable cravings the next day. Instead, I enjoy a giant bowl of salad greens with grilled onions, avocado, bacon, shredded cheddar, and jalapeños, with a juicy, medium-rare wagyu beef patty sizzling on top of it. I relish this meal, eating every delicious bite. After I'm done, I feel like a million bucks, and cravings do not come back the next day!

How does one eat mindfully, you ask? Let me share a few tips.

First, select foods that you enjoy and that make you feel full. If you're not the world's best cook, find someone in the family who is. If you have the time, arrange the foods beautifully—use the fanciest dinnerware you have, even if it is just nice paper plates. The foods you lay out should have a variety of colors, textures, aromas, and temperatures. Play with flavor combinations and unexpected pairings, like a crisp, tart green apple with cheddar cheese. If you enjoy eating with company, share a meal with friends, family, or colleagues. Turn off the television and put down your phone. Chew the food slowly and focus on the taste. *Mmmmm.*

The goal is to make a feast an event with as little distraction as possible, gorgeous presentation, and top-quality foods you love. Then eat until you are full. Do not count the calories, limit the healthy fats, or stop eating because you perceive that others will think you are overeating. Eat until you are satisfied, satiated, and comfortable.

Sound like a dream you will never, ever be able to make true? Sure, I'll admit that what I just described is an ideal situation. In the real world, sometimes we don't have the time, the money, or the inclination to prepare a beautiful meal on fancy china with our loving families smiling next to us at the dinner table. Sometimes we work eleven hours, haven't gone to the grocery store, or our teenager is annoying and won't sit down with us. Or maybe we've fasted for twenty-four hours and are so ready to eat, we shove a hunk of cheddar wrapped in hot bacon into our mouth, burning our tongues and dripping bacon grease over the sink. (Jeez, I hope this isn't just me!)

Situations are not always ideal, but the fact remains that we should both fast and feast. Neither should be a punishment, and both activities can be joyful. We can find freedom in fasting with fewer dishes, less shopping, less cooking, and more time for ourselves. We can find excitement in feasting with foods we love, a new recipe, a dinner shared with friends, and a glass of wine.

The truth is we *deserve* to eat, and we *deserve* to enjoy it. I'll admit that changing what I ate was difficult, and giving up sugar felt like a mourning process to me. But by eating the way I have chosen for myself, I am able to eat something outside my healthy food list 10 percent of the time and still maintain my weight loss.

Your friends and family may also be happy and relieved to see you feasting. Although they'll learn to live with your fasting lifestyle, there are times you're going to complain about being hungry or missing foods that are awful for you (Cheetos, anyone?). When you are fasting, remember that it is your choice to do so, and you can end it at any time. This also means you can complain every once in a while—but never forget that it can get annoying for others. After you have fasted, loved ones are going to *want* to see you feast. They need to know you aren't starving yourself and that you have a healthy attitude about nourishing your body. The best way to prove that to them is to authentically enjoy eating food.

So, feast away. Eat delicious food. Enjoy every bite. Munch on healthy, beautiful, delicious morsels and relish every moment of it. People often say that absence makes the heart grow fonder, and that's true for the mouth, too. Fasting can actually heighten your enjoyment of food!

Eating is a natural human process necessary for the continuation of life. Eating should be a joyful experience. Learning to enjoy food without guilt is a powerful step in learning how to fast.

How to Feast without Guilt

1. Gain control of what, when, where, and for how long you will eat. One hour is an ideal amount of time for eating a meal.
2. Remove distractions from your meal. For example, put away your phone, turn off the TV, and put down the book.
3. Remind yourself that you are a living being who is powered by food and that receiving that food is meant to be an enjoyable experience.
4. Eat until you feel comfortably full. Don't stop eating until you have had enough and don't keep eating because you fear hunger later. There will be opportunity for more food later.
5. Don't compare your food choices and amounts to others'. Each body is unique.
6. Remember that eating is okay. Do not punish yourself for enjoying food. You deserve nourishment no matter your size. It simply isn't natural for humans to loathe doing something they must do to keep themselves alive.

The Problem with Sweets
(and Other Overly Tempting Foods)

I caution against eating foods during your feast that will cause you to gorge or fall into a rabbit hole of addiction. Except for the three times I was in the hospital after bariatric surgery, I ate something sweet every day of my life until I started fasting. Sometimes multiple times a day. Yes, I am addicted to food and, yes, I am most certainly addicted to sweets. I crave them, look forward to them, relish them, and feel excitement from them.

Before I took up fasting, I'd tried to wean myself off sweets, and I'd failed completely. Although I cut down my sugar habit significantly, when I craved something sweet, I'd bake myself a low-carb sweet using erythritol and/or stevia. At first, I found the taste strange, but after a short time, I liked it just fine. I thought, *If I can lose weight and still have sweets—even "fake" sweets—then I can live like this.*

There was a problem, though. Eventually my weight loss stagnated at about twenty-five pounds. I wanted to lose more, and I knew I couldn't do that without taking my diet to the next level. The other problem was that I learned that natural sweeteners like stevia affected my insulin levels even though they have few or no calories. I was insulin resistant, so if I wanted my blood sugar to be stable, the sweets were out. Maybe not for forever, but at least until I reached my goal size and got my health in check.

With my first fast in early 2018, I gave up all the sweetness in my life. It was *devastating.* I got grumpy drinking my morning coffee because I hated how coffee with cream and no sugar tasted. I worried that food wouldn't be exciting anymore because sweets were always my favorite part of a meal. I was like a drug addict detoxing, and I couldn't figure out why, after a fast—when I'd lost weight and felt healthier than I had in years—I could still miss something that was so damaging to my health.

I cheated, a lot. For instance, on my eleven-day fast, I put 4 drops of stevia in my two daily coffees. Once, I was alone at my parents' house, and I shoved the last doughnut in the kitchen in my mouth and threw away the box outside, as if someone was going to conduct a crime scene investigation about the missing doughnut. Another time, I ate a chewable melatonin when I probably didn't need the help sleeping just because I wanted to experience the sweet taste on my tongue for a few moments.

Finally, I decided I *had* to avoid more temptation. I threw out all the sweets in my house, plus my supply of erythritol and stevia. Sure, I felt guilty about being wasteful, but I couldn't have those things around me. I wanted to be healthy. I didn't want a substance to have this much control over me, and I knew, to do that, I had to go cold turkey.

So, I did.

Pretty soon, I started to miss sweets a lot less. I remember one day, when I'd planned to skip breakfast and lunch, I didn't get hungry until 3:00 p.m., when I finally sat down to eat. As I put the first bite of food into my mouth, I realized why I hadn't been hungry: I hadn't been craving sweets.

Today, I refrain from eating sugar and sweeteners about 90 percent of the time. I now like coffee without any sweetener in it, which I swore to everyone would never ever happen. When I do have something sugary, it is usually at a special occasion or a craving I can no longer ignore.

My mind is now clearer, and I've been maintaining a healthier weight for the first time in my life. I'm hardly ever sick, and I feel great in my body. I guess my life is pretty sweet after all.

I'm living proof that you can give up sugar—or any substance you might be addicted to. You may feel like your life will lack excitement without sweets, but as with any addictive substance, it simply isn't true. The sugar is tricking your body and mind into believing that you need it to survive. You don't.

Sugar and sweeteners make it much more difficult for many people to fast because they stimulate cravings. For many, they also stagnate weight loss. Cut your sugar intake in half every week for one month until you get to none and notice the difference in your thoughts, body, and weight. Only after that should you decide how much you would like to indulge.

Feasting Is Not Binge Eating

A controlled feast composed of healthy foods is not the same thing as a binge. While the term *bingeing* is sometimes used casually to describe almost any episode of overindulgence, the reality is that binge eating is a disorder. In fact, it's the most common eating disorder in the United States—and it's one I suffered from until I was thirty-six. During the day I would eat normally, but every night I would sit in front of the television and devour chips, ice cream, meat, and candy—all in mass quantities. The entire time, I'd hear cruel things in my head, telling me, on the one hand, that I was fat and ugly, and on the other, that food would make me feel better. I'd gorge myself until the voices stopped and the stress of my life faded away, covered up by the pain of having eaten way too much.

Abusing food in this way kept me from dealing with the realities of my life. Food shielded me from being happy, sad, and proud and left me an unfeeling zombie, plodding along, trying to do my best without truly experiencing life. I blocked out joy and sorrow with fried chicken and cake.

I was miserable and desperate, but I realized I had to be a positive role model for my young daughter. If I taught her to eat and abuse food this way, I could never forgive myself. So I took action and enrolled in an outpatient treatment facility for forty days. Through various types of therapy, I overcame my binge-eating disorder.

If you think you might be suffering from a binge-eating disorder, rehab might be a good fit for you. But it might not. Many people also benefit from cognitive behavioral therapy. But first I recommend paying attention to your thoughts while you eat. Do they include things like:

"You don't deserve to eat."
"You should be ashamed to eat in front of these skinny people."
"You are the fattest person in the room."
"You are disgusting."

If you feel this way, imagine how you'd react if someone spoke to your daughter, son, best friend, or partner this way. You'd want to protect them from that abuse, right? Then why would you allow yourself to be talked to that way? If you suffer from binge eating, I urge you to seek therapy, where you will learn to speak to yourself with love.

Are You a Binge Eater?

The National Eating Disorders Association defines Binge Eating Disorder (BED) as a severe, life-threatening, and treatable eating disorder characterized by:

- Recurrent episodes of eating large quantities of food (often very quickly and to the point of discomfort).
- Feeling a loss of control during the binge.
- Experiencing shame, distress, or guilt after overeating.
- Regularly using unhealthy compensatory measures (e.g., purging) to counter the binge eating.

If you identify with any of these statements, I encourage you to seek the help of a therapist.

MEGAN RAMOS

Most diets teach people that eating until you're full makes you fat. This misconception is so ingrained in the public consciousness that I've had countless clients (male and female) burst into tears when I've asked them to step on a scale the Monday after Thanksgiving. "I swear I didn't eat any bread, potatoes, or dessert, but I ate *so* much," they always say.

But when the numbers pop up on the scale's screen, their tears of regret turn into tears of laughter. Without fail, almost all my clients have either lost weight or maintained their weight, despite the festive feasting! They just can't understand how they could eat so much food and not gain at least ten pounds!

The Science Behind Feasting

My clients don't gain weight after a Thanksgiving feast because adding pounds is not about how many calories you put in your body versus how many you expend. Weight gain is controlled hormonally, primarily by insulin. Eating lots of refined carbohydrates until you're full will raise your insulin levels substantially, prevent the activation of your body's natural satiety hormones, and cause you to put on fat. But you don't usually eat a carb-heavy meal at Thanksgiving. Sure, there may be bread rolls, stuffing, and pumpkin pie (all carb-heavy foods) laid out on the table, but you typically eat as much or more Brussels sprouts, turkey, and sweet potatoes, which affect insulin much less significantly.

Even if you *did* have a huge carb blowout at Thanksgiving and ate the whole apple pie yourself, are you really in trouble? The answer is no if you "fast it off." I do this all the time. As much as I'd like to tell everyone that I'm some sort of superhuman who never eats carbs, I can't. Occasionally, I enjoy pizza, and I usually eat way too much of it. It's okay. I always schedule a fast afterward. Fasting "fixes" the hormonal blip I got from the pizza by lowering my insulin levels and burning any stored food energy from fat. It also erases any guilt I used to have while eating it, too.

It's tough for many clients to be at peace with eating until they're full, or with eating carbs periodically. All they know is that eating until they're satiated usually results in weight gain, and it typically takes weeks if not months to undo the damage from a big, carb-heavy meal.

This was a tough lesson for me to learn, too.

In 2012, I regained my health and control of my weight with fasting and a low-carb diet, and I made a plan to go to Europe for a month with two of my best friends. Then, something hit me, and I started to cry. How could I go to Italy and not eat pizza or pasta? I wanted to truly enjoy the culture of the countries we visited, and food is so much a part of every way of life. I didn't want to miss out.

Then I realized that my weight, my thoughts about eating, and my desire to feast were completely within my control. If I was planning to eat pizza for dinner or pasta for lunch, I could fast during the day while we were sightseeing. I could also fast during every travel day between cities, just in case I saw a gelato that looked too good to turn down.

I executed my plan on my European adventure, and I lost ten pounds during the time I was away. I even had to buy a new pair of shorts in Rome toward the end of our trip because mine became too big. More than any other time, that trip was when I knew that fasting was the tool that would enable me to control my body and my well-being. I could feast *and* fast and be just fine.

Most of the people I work with have developed fear around eating but are also scared to *not* eat. They've been told that not eating for periods of time will cause them to binge and make them gain weight. I'm here to tell you that feasting and fasting are a natural cycle, and you can enjoy all parts of that cycle. Have faith that eating the right foods or scheduling in a fast here or there around a high-carb indulgence will give you a sense of control. Most of all, enjoy spending time with loved ones during the holidays, and don't worry too much about what your vacation eating habits will do to your waistline. We are *designed* to feast with our loved ones and to fast between celebrations.

Five Ways to Feast Responsibly

1. **PLAN AHEAD**: If you have a holiday or vacation coming up, try to fit in a few fasts before and after the event. These can be short or extended fasts depending on how strong your fasting muscle is. But try to stretch a little. If you're someone who usually fasts for

twenty-four hours, try to fast for thirty-six hours a week before and after.

2. **TIME YOUR FEAST**: It's better to eat an early dinner rather than a late one because you have a few hours before bedtime to burn off whatever you ate. If you're going to eat carbs or sugar, try to do it midday rather than in the evening or nighttime.

3. **STICK TO A SCHEDULE AND DON'T SNACK**: Enjoy your feast while you're doing it and try not to snack before or after it. Sticking within your meal windows is important because it limits the number of times per day your body secretes the fat-storing hormone, insulin.

4. **SKIP THE SODAS**: Try to avoid drinking sugary drinks, especially if you know you're going to be indulging in carbs or dessert at a later meal. Be mindful of what you're mixing your spirits with, too. Margaritas, piña coladas, and many other tropical drinks are extremely sugary, so, if you like mixed drinks, mix the alcohol of your choice with seltzer or soda water and a splash of lime juice instead.

5. **BE MINDFUL ABOUT WINE**: If you're going to drink wine, try to drink a dry wine, which is a wine with no residual sugar. Unlike Moscato and Riesling, which have a higher sugar content, cabernets and dry champagne tend to be lower in sugar so they are a safe choice if you're staring at a wine list and unsure of what to order.

Healthy Eating Habits

Fasting can and will do wonders for your waistline and overall health, but this new pattern of eating is not all you need to maintain your lifestyle. You need a new set of healthy eating habits so you can feast and not get off track forever. These five tips will help you progress in that lifestyle *and* allow you to enjoy the new you.

Tip #1: Spoil Your Appetite

During a meal, it's not just what you eat that matters, it's the order of the food you consume that can make you or break you. I learned this lesson the hard way during a holiday family dinner.

When I first started following a low-carbohydrate diet, I would eat only two of my mother's delicious roasted potatoes at Thanksgiving, and I'd always eat them first. Why? Because I'd been thinking about them for days, and I figured, *It's a holiday, and these are my treats.*

This was a recipe for disaster. The potatoes spiked my blood sugar levels, causing my pancreas to produce a surge of insulin, which made me ravenously hungry. I would end up eating way too much and craving more potatoes, and then I'd feel bloated and terrible for three days after the feast. I thought this meant I could never touch my mother's potatoes again, but it turns out this isn't true. I can still enjoy them, but I need to approach my plate differently.

Now, I eat my food in a different order. I start off with protein, fat, and non-starchy vegetables on my plate. These may include salad, turkey breast, turkey skin, or Brussels sprouts, but, whatever they are, I always eat them before the potatoes. By the time I'm ready for my "treat," I am full and can barely look at the potatoes. At the end of the meal, the potatoes have a very different effect on me. The other foods have already sated my appetite, so I don't crave the yummy potatoes as much.

If you are eating out or going to a big dinner, always start with any non-starchy vegetables like leafy greens, or vegetables that grow above the ground, such as broccoli or cauliflower. Then eat any protein, and save the starches and sugars for last. You will already be full and unable to eat too many of those temptations. You can even go one step further and eat something at home before going out to deliberately "spoil" your appetite. When you are full, it is much easier to avoid tempting foods that may derail your dietary plans.

Tip #2: Don't Eat Mindlessly

Are you eating while standing up? Are you having lunch while working at your computer? Are you eating so quickly that you aren't even chewing your food? You can consume a full meal, but still feel hungry if you are eating mindlessly. We're all busy. Everybody wants an extra six hours in the day, so most of us eat at least one meal a day on the go, which leaves us feeling hungry again within the hour.

Many clients tell me that sometimes they feel full after a meal and sometimes they don't, but they're not quite sure why. I ask them to keep a diary of what they are doing when they eat. These diaries always reveal some version of the same thing: Sometimes my clients have lunch or dinner at the office. Sometimes they are rushing to take their kids to soccer or pick up the grandkids from school. But the story is always the same. When people eat on the go, they're never satisfied. When they take time to enjoy a meal, they experience a strong sense of satiation.

So, what do you do if you don't have time to eat? That's simple. Don't eat. Eat when you have time, and don't eat when you don't have time. Don't eat just to eat. Fasting is always an option. You have the freedom to decide when you consume food. Don't pressure yourself to eat several times a day even if you're not hungry and you're too busy.

Eating when you have time to enjoy your meal means you can go slowly and chew your food thoroughly. Jennifer Lopez claims she chews each piece of food twenty times, so why not you? There is a time lag from when we start eating to feeling satiated, and when we eat too quickly, there is no room for our bodies to register that we have eaten. The result is hunger that just won't go away!

Tip #3: Don't Go Food Shopping When You're Hungry

Grocery shopping while fasting usually means lots of regret the next day. If we're hungry in the store, we're much more likely to gravitate toward the junk foods we're trying so hard to resist.

I try to go to the farmer's market or the store on a Sunday morning when I'm not in a rush. This has really improved my relationship with food and the quality of foods that I buy. I've noticed that if I'm in a rush, I'm more likely to buy "quick and easy" refined foods that I'll later regret.

Tip #4: Eat from Smaller Plates

When I was creating my wedding registry, I told the store associate, "I want the smallest dinner plates in the store!" She looked at me like I had ten eyeballs and said that it was the first time anyone had ever asked her that. Every other customer she'd helped wanted the largest plates possible. And apparently, the store had figured this out, too, since they only had one normal-size plate and two dozen sombrero-size ones.

When you use a smaller plate, it sends a signal to your brain that you need to eat less. This is because adults tend to use external cues to figure out when to put down their forks. This is the opposite of children, who rely mostly on internal cues to signal them to stop. If we finish the food on our smaller plate, we can always get more. But a small plate forces us to think carefully about whether we want more. It helps us eat mindfully. Are we eating because we are hungry, or are we eating because we want to finish our plate?

Tip #5: Only Eat at Mealtimes

Even with "healthy" foods, snacking is the quickest way to start packing on the pounds. A handful of almonds here and there, and pretty soon the weight starts coming back. Almonds are fine, but make sure they're part of your mealtime. Remember: snacking is not your friend.

Some practical strategies for implementing non-snacking may be to allow yourself to eat only at your dining table. This will help break the habit of mindless eating, which we shouldn't be doing. We should be eating only when we're hungry and not if we aren't. Are you snacking because you are hungry, or are you doing it because you are bored?

I used to have a long commute, and I would often buy a bag of nuts at the gas station. This was a hard habit to break until I had my car professionally detailed, and the man who cleaned the inside said he found a cup of almonds scattered throughout it. When I got a new car several days before a road trip from Toronto to Orlando, I was so determined to keep it clean and food-free that I ended up fasting for the entire car ride! So, take some time to tidy up your car, giving yourself a "clean slate" and reducing your snacking on the road.

Another hack I recommend to people who watch TV in the evening is to buy a puzzle book and keep it on your coffee table. This way you can distract yourself during commercials—which are often about junk foods—with something that will exercise your brain.

Above all, don't be afraid. You *can* enjoy food between fasts, and you *can* feast appropriately. Between these tips and using fasting as a tool when necessary, you'll be able to maintain your new weight and be happy and healthy for years to come.

CHAPTER 17

Meeting Your Goals and Going the Extra Mile

EVE MAYER

By now, you may have completed several longer fasts and made intermittent fasting a part of your lifestyle. Everyone reaches their goals at a different pace, so if you haven't hit them, that's fine. You'll get there!

Tracking Your Success

As you progress with fasting, you're probably going to want to measure your success. Tracking how far you've come is about so much more than pounds, though. Your success can be measured in the activities you can now do, the medications you don't have to take, or the hours you're sleeping after years of waking up due to health complaints. Remember those personal goals you wrote down before you started fasting? Pull out your goal list and look at it. And, if you didn't write down your goals, quickly think of some and write them down. Right now. It's okay, I'll wait.

Ready?

You can now adapt that list by creating your own goal tracker. On a piece of paper, online, or on your phone, make note of the date you started fasting, where you are now, and another date you consider a goalpost. This "goalpost" date could be your birthday, your wedding day, or any random day. It also doesn't have to be an end date. After all, fasting is a lifestyle, not a race, and I encourage you to make it part of your life forever.

Here's a goal tracker I used. My goal is to get to 29 percent body fat. I am currently at 36 percent body fat, so I wrote down the day I made that goal, a day by which I'd like to reach that goal, and a few goalposts in between. If you're tech savvy, you can get fancy and make your tracker a spreadsheet that automatically graphs, or you can be lazy like me and sketch it out. Regardless, you should use whatever gives you a motivating visualization of your process toward your goal.

For example:

Start October 2018	Goalpost #1	Goal!
36% body fat	32.5% body fat	29% body fat

I've found that this is the simplest way to track my progress toward my goal, and it's one that focuses my attention on my overall health and happiness rather than just my weight.

• • •

Are Scales for You?

To weigh or not to weigh. That is the question. Just like you must make the decision of what to eat and when to fast, you must also decide if getting on the scale helps or hinders you. My husband loves the scale, and the first thing he does in the morning after going to the bathroom is to step onto it. Sometimes it goes up, and sometimes it goes down, but either way, he stays levelheaded and logs the data point into an app. His brain processes the number on the scale as data, or simple information that can be changed by pulling one lever or another. He feels that weight is meant to fluctuate over time, and the data never affects his mood. He is what I might refer to as scientific and reasonable, but he is *not* like me. If you relate to these feelings of levelheadedness and find that weighing yourself every day gives you information that helps you make dietary and lifestyle decisions, then go for it.

When I weigh myself, I respond very differently than my husband. I am sensitive (to say the least) and when the scale goes up, I immediately experience a flurry of emotions including guilt, dread, sorrow, regret, shame, sadness, and fear that I will become fat again at any moment. These emotions happen whether I gain 0.3 pounds or 3 pounds in one day. When the scale throws an ugly number in my face, I want to cry—and, sometimes I do, especially if the efforts I have made in the days before do not match up with what I see on the scale. A number I dislike on the scale can very well set the tone for a bad day for me no matter *what* happens.

Then there are the days that I step on the scale and the number is down from the day before. My emotions include glee, joy, excitement, expectations of grandeur, superiority, happiness, giddiness, and fabulousness. If the scale goes down after a day that

I enjoyed the very rare indulgence of a birthday cupcake, I feel *victorious*. My mind says, *Ha ha, scale. I have finally fooled you. Yesterday, I ate an entire cupcake with tons of icing, and you didn't even figure it out. I can probably eat whatever I want from now on because even when I cheat, I don't gain weight. Finally, I have figured out how to trick my body!*

Trust me, I'm not proud of these thoughts, and I know they're not based on any kind of reality. But I now understand my food-obsessed mind, so I choose to not weigh myself every day for the sake of my sanity.

The fact remains, though, that the scale doesn't always make sense. When I completed my first extended fast, I lost weight every single day until suddenly I started gaining weight back toward the end. That's right. I hadn't eaten a morsel of food for over a week, but I still packed on pounds. I also weighed 195 five years ago and could barely zip my tight jeans if I wanted to enjoy the sport of breathing. When I arrived at 195 again this year, those same jeans were sliding off. Why? Because I had built up more muscle and carried less fat. My body weighed the same, but I had a healthier, more balanced composition, with more muscle.

The scale method doesn't work for everyone, nor does it take into account everything, like the fact that muscle is more dense than fat. If you've been weight training, your clothes may fit better (like my jeans did), but you may weigh more than you did before you ever hit the gym. The scale also doesn't account for other very important indicators of health, like bone mass, blood pressure, heart rate, and more.

If you're on the fence about whether to use the scale, here's a way to gauge your dependence on it, as well as your emotional reaction to it: Weigh yourself every day for one week at the same time each day. Record your weight and your feelings. If you're not alarmed by weight fluctuations, keep doing this. If you panic every time you go up an ounce, maybe you should weigh yourself once a week, once a month, or not at all. Pick an interval or time that suits you best, and if you are too obsessive about weighing all the time, donate the scale to a friend or Goodwill. If you have family members who want to keep the scale, then have them hide it from you.

Tracking Your Measurements

Instead of getting on the scale every day, I prefer to take my measurements, including my body fat percentage and inches lost or gained. Bear in mind that these measurements don't fluctuate as rapidly as weight, so taking them daily—or even weekly—can be problematic. They can also differ greatly based on who is taking them and what instruments are used. For example, mine are totally different depending on whether I take the measurements, my husband does, or my trainer does. Try recording your measurements once a month, and have the same person use the same tools to do them.

I'm pretty old school and use a tape measure to record my inches gained or lost. But if you want to go high-tech and don't mind spending the money, I find DEXA body composition scans to be very effective. You can search online in your area and find places that specialize in testing this way. There are usually packages you can buy to reduce costs if you want to go back monthly or quarterly, which is what I do. These machines are quick, painless, and give you a full report of your measurements, body fat percentage versus lean mass, bone mass density, and more.

DEXA scans are very simple—you remain fully clothed while

a 3-D scanner analyzes your body. When the results come back in minutes, they include graphs and data, and if you return later for another scan, you'll receive graphs that compare the numbers from your earlier visit.

These scans show me things I had never thought about before. For example, apparently I have very good bone mass. Why? Probably because I was super fat for twenty years, and my bones got very strong carrying all that weight around. They also show that my visceral fat (the fat around the internal organs) is decreasing, which is great news because visceral fat is an indicator of a propensity to develop specific health problems like heart disease, metabolic syndrome, or type 2 diabetes.

Bear in mind, however, that DEXA scans will show you a side of yourself you might not want to see! Before my first scan, I was left alone in a room with directions to strip down as much as I felt comfortable. I was told to stand on a circular platform while holding two handles, then let go and press a few buttons. I wanted the most accurate measurements possible, so I got totally naked. The technicians had directed me to put my hair up, so I piled it on top of my head like a sumo wrestler. With my feet hip width apart and my naked body exposed to the machine, I held on to the handles and pressed the buttons as the circular platform began to spin me around slowly. I now understand exactly what a rotisserie chicken feels like!

When I put my clothes back on and left the room, the staff said the information would be delivered to me via email that day. I left the office, and when I returned home, I opened my email. There it was! Excited to receive the data, measurements, and information about my body, I moved the mouse to open it, but was *shocked* to see what was in the email's preview bar. What was it? Motion pictures of my naked, rotisserie chicken–like body spinning around on my extra-large monitor screen. Sure, it was a computerized image, but anyone could have been able to tell it was me. It was shocking, devastating, and I was not prepared to see every flaw on my body. I thought I was already well acquainted with my many imperfections, but I was wrong!

After the shock wore off, I became intrigued by the spinning statue model of my body, and now I find these types of scans useful for accurate measurements. But I could have used a word of warning that my naked body would show up on email!

Other Ways to Track Your Success

You can measure your success in any way that works for you, but here are a few methods you might want to try:

- Wear the same outfit on the same day each month and have someone take a full-length picture of you in the same spot. Put the picture on the fridge each month with your date and weight on it.
- Find a pair of jeans you can't fit into anymore and try them on once a week.
- Wrap string around your waist each week and cut it down to the size of your waist. Hang the string each week starting at the left and watch as the strings get shorter and shorter— sometimes even when you don't lose pounds on the scale!
- Monitor your blood sugar and use an app to track it. Review your progress weekly.
- Take your blood pressure twice a day with an automatic blood pressure cuff—available online—and use a spreadsheet to track it. Render a graph each week to see your progress.

- Walk around the block and time yourself. Track the time once a month and compare it to last month's.
- Do a timed plank once a day and track how long you can hold it.
- Take a selfie at the same angle once a month and compare it to the previous selfies.
- Remember what hole in your belt you are on and see if it changes.
- Notice if you are out of breath when you walk up a set of stairs.
- Go to your doctor for a full blood panel, including cholesterol, lipids, and triglycerides. These are often fully paid for by insurance during an annual physical.

Measuring your success is essential to achieving it. You can't know that you have succeeded if you don't know what you are trying to accomplish—or how far you've gone. Sometimes you will hit a dry spell where it seems like forever since you have had a win. Other times the wins will come one after another in bursts. Track your measurements and bask in your success. Share them with someone who cares about you and supports your goals. We spend so much of our time beating ourselves up, but it is time to lift yourself up and give yourself credit where credit is due.

Once You've Reached Your Goal

You did it! Congratulations, you're a rock star!

Now you have a choice to make. You can decide that your original goal is enough. This means the goal you started with is where you want to stay. Or you can decide to move the goalposts a little further and settle on a more aggressive goal—one of less weight, greater health benefits, or more life rewards. No matter what you decide, it's up to you, and you can stop at any point. After all, a lady or a gentleman *always* has the right to change their mind.

The truth is that it is perfectly okay to be happy with where you are now. If you find yourself in a body you're content with, and your health is good, there is no need to push further. Instead, notice yourself, congratulate yourself for what you've accomplished, enjoy it, and learn what you need to do to maintain it. Make sure to give yourself some credit and to find an encouraging way to monitor that you are staying on track. Remember that your goal does not need to be a number on a scale. Your goal can be increased flexibility, fitting into an old pair of jeans, a specific body fat percentage, the ability to hike without wheezing, or whatever you deem worthy of your effort.

Just bear in mind that some people find it as difficult, or more difficult, to maintain their goal as they did to get there in the first place. I'm not saying this to scare you; I'm just being honest. Settling into a place where you're happy with your goals doesn't mean giving up and ordering the super-size cheeseburger, fries, and soda meal at your local greasy spoon, or deciding never to fast again. You have to keep working, even when you're maintaining. But, remember, you may slip up or fail from time to time, and that's okay. In chapter 21, we'll discuss the many ways you can get yourself back on track.

Making Bigger Goals

Once you have properly celebrated reaching your goal and maintained it for the amount of time you wish, you may decide that you want to go the extra mile. This means you need to set a brand-new goal. For example, I am currently maintaining the achievement of my first body fat percentage goal—37 percent—but when I decide to start another race with myself, I'm going to push for a body fat percentage of 29.9 percent. The same rules apply for setting your goal that you used the first time. If you need a refresher, turn to chapter 7: Ready, Set, Goal!

Deciding that reaching my goal is enough, and that I don't have to keep trying, is a strange experience for me. I'm an overachiever, so my impulse is to keep working until I collapse. But, today, I am learning to give myself credit for my accomplishments, enjoy my new body, try out new clothes, hike for the first time ever, and accept that what I always wanted finally happened. I wish the same for you!

When (Good) Accidents Happen

Just bear in mind that your decision to change your goals may come by accident. When I reached my most recent weight-loss goal of 195 pounds, I was ecstatic. After a few decades of losing weight and gaining it back, it was the first time I was able to stay there for a week. Then, through a combination of fasting and keto, I started losing even more weight. I ended up at 185 pounds, and, for the past seven months, I have bounced between 181 and 189. This feels like a stellar achievement to me, and for now, it is enough. I am filled with pride and am still adjusting to the fact that I am in the 180s and shop in regular stores. I honestly didn't believe I had a chance at a life that didn't include shopping in plus-size stores and ordering wide-size shoes! Plus, I feel incredible. I've only been sick once this year, which is a jaw-dropping improvement from getting sick five or six times a year.

Moving Ahead

Setting a second or third or fourth goal does not diminish your achievement of your first goal. This is like reaching the moon and deciding to visit Mars. Why not? You can decide what you want, and the journey can end or continue as you wish. My only suggestion is not to have more than two goals max running concurrently. This helps you to focus clearly and visualize your eminent success. We as humans aren't great at working toward many things at once. There are always new goals to achieve, so you never have to worry about running out, and you can always strive to be your best self.

PART IV

Problem Solve
Your Fast

CHAPTER 18

Solving Health Complaints

MEGAN RAMOS

It's reasonable to expect that you'll experience some new physical changes while you're fasting. Yes, you'll need to learn how to deal with hunger, but you might also experience unexpected issues like increased thirst, headaches, or sleeplessness. These symptoms don't happen to everyone, but they do occur, and they're nothing to be concerned about. If you encounter more serious health issues like anemia, kidney or liver dysfunction, or irregular heartbeat—all of which are rare—you need to seek medical advice.

In this chapter, we've compiled a list of some of the most common physical reactions people experience when they begin fasting. Please bear in mind that everyone is different, and everyone's body responds differently to fasting. Again, if you experience more serious symptoms, it's best to consult with your doctor.

• • •

Bad Breath or Bad Taste in Your Mouth

People sometimes report a metallic taste in their mouth when they start fasting. Others report that their breath smells like nail polish or fruit. This is a sign that your body has entered ketosis. Ketones—beta-hydroxybutyrate, acetoacetate, and acetone—are expelled through urine and exhalation, and they may have an odor (for instance, acetone is one of the ingredients in nail polish). This odor or taste will go away over time, though you may need to brush your teeth more often for the first few weeks.

Bloating

If you're bloated, you may have consumed too much salt, and your body is retaining water because of it. Try reducing your salt intake or drinking water rather than bone broth or another salty beverage.

Coldness

If you feel cold while fasting, that's a sign that your body is entering ketosis. Your body is simply having a bit of trouble moving from burning glucose to burning adequate fat to keep you warm. It's nothing to be concerned about. You should warm up once your body becomes fat-adapted and fully makes the switch from burning sugar to burning fat.

Constipation

Fasting causes your insulin levels to drop, which sends a signal to your kidneys to release stored water. This can cause you to become dehydrated, then constipated. During the days that you aren't fasting, increase the amount of leafy greens and fiber you consume. During your fast, soak in Epsom salt baths and take magnesium citrate (at a starting dose of 400 mg once per day) if you must. Hydrate with salt and water.

Many people assume they're constipated during an extended fast when, in fact, they just don't have anything in their intestines. If you haven't had a bowel movement during a long fast, but you aren't cramping, then you're fine.

Depression

As we discussed in chapter 1, people often report feeling less anxiety and experiencing an improved mood when they fast. Depression is unusual, so if you do feel depressed, we suggest seeking the advice of a therapist.

Diarrhea

If you experience diarrhea while fasting, try mixing 1 to 2 tablespoons of chia seeds or psyllium husk with water, wait 10 minutes, and drink the mixture. Chia and psyllium both absorb excess water in the digestive tract, allowing less liquid to be expelled with loose stools.

Dizziness

Dizziness is often a sign of mild dehydration. Be sure you're drinking water throughout the day. Consume extra salt in the form of pickle juice or bone broth. Some people, like Eve's dad, may also be experiencing a stabilization of their previously high blood pressure. If you're on blood pressure medication, have started fasting, and are experiencing dizziness, be sure to see your doctor. You may need to lower your medication dosage.

Dry Lips

This may sound counterintuitive, but dry lips are a sign you are drinking too much water and not enough salt. Add salt to your diet in the form of pickle juice or bone broth.

Fatigue

When you start fasting, you may feel tired at first as your body transitions from burning sugar to burning fat. This should pass within your first three or four fasts, when you begin to feel more energetic.

Headaches

Headaches are very common when you're fasting. We don't exactly know why they happen, but there is speculation that they may be due to a salt deficiency. To prevent and treat headaches, consume more salt in the form of bone broth or pickle juice. Stay away from ibuprofen and other painkillers.

Heartburn

If there's no food in your belly to absorb your stomach acid, some of it may come up, causing a tight, burning sensation in your chest. Try an over-the-counter antacid or speak to your doctor about a prescription if the problem persists.

Intense Emotions

Volatility and feeling intense, often explosive emotions aren't incredibly common, but they may happen when you start fasting. It's a dramatically new lifestyle, and it's normal for your mind to feel overwhelmed. Hang in there! Surround yourself with people who love and understand you, and seek professional therapy if needed.

Nausea

Feeling nauseated is not considered normal while you're fasting, and it can be a sign of dehydration. Drink water, and if you experience anything more than short bursts of nausea, or if nausea increases to a degree that makes you uncomfortable, stop fasting.

Sleeplessness

Most people report having sleep issues during their first two weeks of intermittent fasting. This is the result of your body adapting to the increased level of adrenaline that it produces while you fast. Create a calming, relaxing bedtime ritual to try to wind down. You might dim the lights, drink a cup of warm herbal tea, and curl up with a book until your body tells you it's ready to sleep. Avoid looking at screens—televisions, phones, laptops, tablets—before bed, as they emit blue light that interferes with sleep. If sleeplessness persists, try taking a melatonin supplement.

Thirst

Increased thirst is completely normal while you're fasting because after your body burns the fuel in your stomach, it burns glycogen. Glycogen is bound with water molecules. Ever hear someone say that, instead of losing fat, you've lost water weight? That's what they're referring to. If you feel thirsty while you're fasting, drink more water. Strive to consume at least half your body weight in pounds to ounces and drink that each day.

Upset Stomach

An upset stomach—as opposed to nausea—is likely just the result of hunger pangs, which can be alleviated by drinking mineral water.

CHAPTER 19

Mind Tricks
While Fasting

EVE MAYER

When I started fasting, I thought that I would be in a constant battle with my stomach, but I was completely wrong. For me, the biggest struggle was with my mind.

Before fasting, my life centered on food. I planned meals, anticipated my cravings, shopped, ate, cooked, researched restaurants, made reservations, ate, took pictures of my food, posted those photos on social media, ate, came up with clever food hashtags, discussed the food with friends, cleaned dishes, wrote food reviews, and ate some more.

Then I started fasting, and I realized I had to center my life on something besides food. But to do that, I needed to retrain my brain.

• • •

Battling Boredom

When I began to practice intermittent fasting, I found that I suddenly had a staggering amount of free time. I had hours with nothing to do, and I got bored—fast. I needed to develop new hobbies and interests that would take my mind away from its habitual desire to snack or sit down to a meal. I started mentoring, shopping with my daughter, and reading more books. While it took a few weeks to find my groove, now I can't believe I ever preferred spending my time sitting in front of the TV, shoving food in my mouth!

It may be so long since you've cultivated a habit that you find it difficult to get started doing *anything*. Don't stress out about that, especially since anxiety might cause you to want to snack. Instead, try to remember the things you enjoy doing with friends and family—other than eating. If you can't recall any, ask them or, even better, find out what kinds of activities (other than eating!) your loved ones would like to do with you. Often people are surprised to hear that their kids have missed spending time with them because all those hours had involved being stuck in the kitchen. Start by going on a walk with your family or buying tickets to a concert. You may discover a love of live music or the outdoors you'd long forgotten.

Before I started fasting, I found it helpful to make a list of the things I enjoyed doing outside of eating. I was shocked to remember how much I liked throwing the Frisbee to my dog, walking in my neighborhood, reading, reaching out to old friends, and organizing my house. Early on, when my mind was wandering, thinking about how much I'd love to drown my emotions in a tub of ice cream, it was therapeutic to revisit this list. Like I said, it took me some time to reengage with many of my old interests, but now I can't imagine my life without them.

Forty Things to Do Instead of Eating

When you're fasting and bored, you don't have to do something strenuous, brain-busting, or taxing to fill up your time. You can pick from any of these activities, which will help the minutes (and maybe hours!) fly by.

1. Drink water.
2. Listen to music.
3. Call your mom.
4. Research your next vacation.
5. Go for a walk outside.
6. Clean the kitchen.
7. Walk your dog.
8. Read a book.
9. Update your résumé.
10. Call a friend.
11. Visit the library.
12. Knit or crochet.
13. Journal.
14. Watch a TV show you've never seen before.
15. Do ten jumping jacks.
16. Make a cup of tea.
17. Fold laundry and put it away.
18. Go to a yoga class.
19. Go shopping.
20. Clean out your email in-box.
21. Visit a neighbor.

22. Drop off the dry cleaning.
23. Say a prayer or meditate.
24. Organize a closet.
25. Sit outside in the sun.
26. Take your child on a fun adventure.
27. Organize your digital photos.
28. Scrapbook.
29. Play a board game.
30. Listen to an audiobook.
31. Garden.
32. Make a craft project.
33. Go to a museum.
34. Write a thank-you note.
35. Read a magazine or newspaper.
36. Tackle the junk drawer.
37. Watch old family movies.
38. Clean the bathroom.
39. Pick wildflowers and put them in a vase.
40. Rake or sweep outside.

Things Not to Do While Fasting

There are so many activities that you might not *think* are centered on food, but in fact highlight food prominently. If you participate in them while you're fasting, I can promise you you'll start doing mental somersaults around the idea of eating, and before you know it, your hand will be in the cookie jar.

I'm warning you! From the moment you begin a fast to the minute you end it, you should avoid the following activities:

Looking at Social Media

Guess what people post on social media? Pictures of delicious-looking food! If you are like me, when you are first learning to fast, it's impossible to resist running to the refrigerator when images of food pop up every two seconds. Worse than the temptation, though, is the self-loathing you might feel. I've found myself getting frustrated looking at these photos and associating them with the person who posted the images. I start to think, *How can [this person] eat all this sugar and fried food and look so great? Why am I the only person who can't have food?*

Of course, these are ridiculous statements, but they eat away at you at a time when you need to be strong and confident. Remember, you are in control. You are the person who has made the decision to get healthy. You are the one who has decided to fast, and *you* can stop fasting at absolutely any moment.

Grocery Shopping

Before I started fasting, I'd go to the grocery store determined to get lettuce and tomatoes, but instead return home with Cocoa Krispies and Pop-Tarts. I'd shop while hungry and snack on something straight out of the bag. I rarely had a list or a plan when I went to the store, so it became a free-for-all involving me grabbing random ingredients hoping I could come up with enough dishes to make for my family. This often led to repeat grocery trips each week, wasting my time and money. My lack of planning always led to me throwing out food that had gone bad, which made me feel guilty for being so wasteful.

Fasting helped me become a better grocery shopper and a better cook. If I was going to eat fewer times, then I was going to make sure the meals I did prepare were delicious and enjoyable.

When I'm fasting, though, I can't grocery shop. Even if I'm not hungry, I start battling with my reflexes any time I see a free food sample in the grocery store. As I push my shopping cart down the aisles, I feel like I am surrounded by temptation and things I can't have. I become so resentful and deprived that I usually break my fast sooner than I want to. Sometimes, I go on to eat more than I truly need.

I strongly recommend planning your grocery shopping carefully before or after your fast, even if you are feeding a family. Ask for help from family members or use a delivery service if possible.

Cooking

When I am fasting, I do not cook. The sight, feel, and smell of food are too much for my senses to overcome. This is not about my lack of willpower, either. When your senses are bombarded by food, it's only natural for you to want to taste it.

I am fortunate to have a husband and an eleven-year-old daughter who can cook for themselves. Not everyone is in this position, though, so if you have to cook while you're fasting, here are a few tips I recommend:

- Prep your family's meals before your fast so you can just heat things up.
- Ask for help from family members who are able to cook.
- Buy prepared foods or send your family out to eat.
- Cook items that are their favorites but aren't yours.
- Turn on the vent or open the window to reduce the smell of food.

Cleaning Dishes

You might be surprised to hear this, but cleaning dishes for others is tough for me when I am fasting. Years ago I "accidentally" stuck my finger in leftover pudding and licked it, and once a piece of chicken on a dirty plate ended up in my mouth on hour forty-two of a forty-eight-hour fast. I had decades of experience snacking and nibbling, and breaking this bad habit took time.

Now, I am much more accustomed to fasting, so I could probably wash dirty dishes with ease while I'm not eating. However, I'm not going to mention it to my husband because I hate washing dishes and he is kind enough to always take care of it. (Sssshhhh!)

Going to the Mall

Going shopping at the local mall seemed like a great idea when I first started fasting. After all, what was I supposed to do with all that extra time on my hands? Indulge in a little retail therapy! But when I stepped into the mall, the smell from the pretzel shop wafted out and slapped me across the face. I felt angry at the people walking around holding cinnamon rolls and pizza slices, and I wondered if I could just maybe have a bite or two.

Today, I can go to the mall with ease while fasting and still love my fellow humans—even when they have a giant pretzel. But that wasn't the case at first, so don't make the same mistake I did.

Going to the Movies

I love going to the movies. It used to mean a cool, safe place to shove all the popcorn and candy into my face in the dark without the judgment of others. As I changed my eating, I switched the popcorn and Coke out for a giant dill pickle and water, and I felt pretty happy. But, even today, the idea of fasting at the movies just does not sound fun to me. I know I should be well adjusted to fasting and at the theater only for the entertainment, but, unfortunately, I can't break the habit of wanting to eat in a theater. You may feel the same.

I suggest you remove yourself from the situation and plan your movie theater outings for when you're feasting. You can watch plenty of movies at home without needing to snack.

Going to a Party

I plan my fasting around social events. There is a time to fast and there is a time to feast, and, for me, a party is a time to feast.

As a middle-aged married woman with a preteen, I realize that my social life isn't filled with fancy, food-centric soirees all the time, so this is easier for me. If your lifestyle requires frequent social gatherings, and you can't schedule your fasts around them, try holding a glass of water or a club soda with lime in your hand while you're working the room. People will question you less about eating, and having a glass to carry will give you something to do. In addition, waiters will offer you food less often if you are drinking and socializing. You may think people watch everything you eat, but often they won't even notice if you go to a party, have a few cleverly disguised drinks, socialize all night, and eat nothing. If they do offer you something to eat, the easiest thing to do is to thank them and say you aren't hungry at the moment.

Going on Vacation

I am comfortable with intermittent fasting on vacation, but extended fasting is tough for me if I'm away from home. I like to experience different cultures and places through food, so a long fast while I'm on a trip makes me feel like I'm losing opportunities to learn something. However, intermittent fasting on vacation works great for me because I feel like I can make one or two meals that day really count with a feast.

• • •

This list of dos and don'ts while fasting may be different for everyone. Like anything else, fasting truly gets easier with practice. I honestly didn't believe that when I started, but it is true. At first, I couldn't go to parties, the mall, or near any store with food while fasting. Now, I can do most of these things with ease. Just remember: You get to make the decisions of when you will fast and when you will stop. You get to be in control. You get to decide what is best for your mind and body during this entire process. For control freaks like me, once you accept that you are truly in charge, fasting becomes an enjoyable thing to do.

Finally, don't judge yourself so harshly if you need special circumstances when you first start fasting. I made that mistake and saw myself as weak.

I wasn't weak, I was inexperienced, and I got more experienced over time.

Now, I know I am strong. I gave myself the time and the situations to develop the skill of fasting. I am not the secret loser I thought I was! You deserve to give yourself the best chance of success. And bit by bit, you'll see your body and your mind begin to change into what you always dreamed they could be.

JASON FUNG

I've understood since I was a kid that the mental challenge of fasting is far more difficult than the physical challenges. Physical hunger passes very quickly, and, luckily, I learned this firsthand many years ago.

When I was twelve years old, I had to undergo a gastroscopy and colonoscopy because I'd become anemic. Doctors eventually discovered I was suffering from peptic ulcer disease, but before my diagnosis, my symptoms were unexpected and mysterious. During a gastroscopy, a small camera is inserted into the mouth and passed into the stomach to search for bleeding. In a colonoscopy, a tube is also inserted from the bottom up, to look at the intestines. Before the procedures, I wasn't allowed to eat anything for twenty-four hours. On top of their concern over what I might be suffering from, my parents were worried that I'd be ravenously hungry and become weak. After all, kids are growing, and they need food—and a lot of it!

But what I remember more vividly than either procedure was how normal I felt during the day I didn't eat. I wasn't tired, I didn't go crazy, and I didn't pass out from hunger. I was mildly uncomfortable at worst. I was shocked, but I filed it away for years until I began to recommend fasting to my clients.

Why did I not think about fasting for years? Because we live in a culture that values and promotes nonstop eating. When you're fasting and run headlong into well-meaning people who don't understand that fasting is, in fact, totally healthy and natural, I recommend employing some mind tricks to turn that thinking on its head.

Know That Fasting Is Common

There are societies where millions of people fast on a regular basis. Many Muslims fast during the holy month of Ramadan. Many Catholics fast during the season of Lent. Many Jews fast during Yom Kippur. Buddhists, Mormons, and Hindus all have their own fasting traditions. The point is not the specific traditions of each particular faith but rather that, within these societies, it's not strange for large groups of people to fast on schedule more or less throughout their entire adult lives. They seem to do it without much difficulty, it's not frowned upon, and even their mothers don't worry that they're going to pass out, stop growing, or die from hunger. Yet in our society today, most people I approach with the idea of skipping even a single meal seem dead certain that they will be quite unable to do such a thing. They think that not eating even a single meal is impossible, is strange, or breaks all the rules of civilized society.

When you embark upon a fast and struggle with the idea that you're doing something strange or unprecedented, don't worry. You're not alone, and you're not doing anything wrong! Your choice is totally normal, and billions of people throughout history have done exactly what you're doing on a regular basis.

CHAPTER 20

Fasting with a Social Life

EVE MAYER

Now that you're fasting, you may fear that it's time to put your party clothes in storage. Not to worry: fasting doesn't mean you have to give up your social life, vacations, celebration time with family, or anything else that's important to you. This chapter will guide you through the thorny logistical and personal issues surrounding your social life. I promise: a successful fasting plan and a robust social life are not mutually exclusive goals!

Scheduling Is Everything

Before we get into the nitty-gritty of how to navigate social situations while fasting, you need to understand some basics. One of those involves getting organized. While fasting can be incredibly flexible, working for all kinds of people in all kinds of situations, one critical component of any type of fast is sticking to a pre-planned schedule.

At this point I've been fasting for so long that sometimes I allow myself to become *too* flexible with my approach. I may think to myself, *Hmmm, I haven't been eating the way I should for the past couple of days, so I'll just fast tomorrow.* Then I push back tomorrow a couple more times, and suddenly I don't fast for a week. This usually results in a one- to three-pound weight gain for me over one week. And it isn't muscle I put on during that period. It is *fat.*

On the other hand, my obsessive tendencies can get the best of me. I may start a fast that I intend to do for twenty-four hours. The fast will go smoothly, and I'll start to think, *Hmmm, why not extend this to thirty-six hours, or forty-eight hours, or maybe three days, or a week?* Suddenly, all my well-laid plans about work and family are up in the air, and my life feels more disorganized than it's ever been.

These examples are meant to illustrate that fasting is so flexible that you can easily fast too much or not enough, thinking that you can just adjust your life accordingly. If you're like me, you can't; life is just too busy not to stick to a plan. Unfortunately, no one can give you a prescription for the exact schedule you need. It's up to you to figure that out.

When I started fasting, I found this incredibly maddening. I figured that if I found the best specialists in the world, they could just give me the magical fasting schedule that would allow me to meet my goals. But though I met Dr. Fung and Megan Ramos—the best in the business—early on in my fasting journey, even they couldn't offer such a magical solution.

Now I get it. Megan and Dr. Fung are such pros that they know an ideal fasting schedule must be discovered by people for themselves. Experts can guide, suggest, and share what has worked for thousands—and, in fact, Megan will offer a few great sample plans starting on page 218—but they cannot know the particulars of your life. They can't know that you love to eat late at night or that you cook for three kids every day or that when you pass Mc-Donald's your car magically turns into the drive-through. They won't know that Sunday brunch after church is sacred in your family, and that you need to plan your fasting schedule around it, or that you have a business dinner planned for next week. We all have our quirks, triggers, and schedules to overcome, and only you hold the key to figuring those out.

Fasting requires balance, accountability, and a plan. You get to pick this schedule, and it is customizable to your lifestyle, but you need to work hard on it and adhere to it. If you love to make schedules, then this will come easily to you. Once you decide how often you want to fast on a regular basis, take a look at your life and fit fasting into it.

Here are a few tips on fitting fasting into a wonderful social life.

Don't Be a Martyr or a Jerk

When I began fasting, I did not know how to plan my fasting around my life. I didn't strategically select when I should fast, nor did I think about how I'd feel when I fasted. Therefore, when I went to work and saw the free breakfast or lunch my company supplied each week, I'd glare at my co-workers. Sure, I knew when both the free breakfast and free lunch were scheduled because they were on the same day every week, but I thought I'd be okay. I wasn't. Even though my co-workers were supportive of me the whole time I was fasting, I was an irritable, resentful, all-around-grouchy colleague.

You can learn from my mistakes.

I realize now that it was acceptable to announce my weight loss in the portion of our Monday meeting devoted to each person's personal and professional win of the week. I also know that it would have been perfectly okay for me not to tell anyone I was fasting. It's my business, and I could have avoided a lot of questions if I'd done that. However, it was *not* acceptable for me to fast during the free company meals and be resentful of my colleagues. It was *not* collegial of me to complain in public about not being able to eat breakfast, nor to announce loudly that I couldn't have lunch because I was fasting. It was disruptive to my colleagues to wear my terrible circumstances like a badge of honor and bitch about wishing I could eat.

I acted like a martyr, and I regret it.

Luckily, it took only a few months for me to learn how to turn this behavior around, but I wish someone would have told me from the beginning how to not be a jerk when I was fasting. So here it is: When you fast, don't make it all about you. Fasting is within your control, and you've chosen how and when to schedule it. Don't act as though you're helpless or suffering because you decided to do it over Thanksgiving, a vacation, or on free breakfast Monday at work. *You* are in charge of your fasting schedule, so own it.

When to Fast

If possible, plan your fasting to avoid group activities where food is free, available, or everywhere. I could have easily chosen not to fast on the days when my company offered free meals, allowing myself to make healthy food choices with them (or sneak in a doughnut) and fast during the times when food would be less central. I wasn't fasting every day, so I could have been flexible.

If you are going to be at a group event or party where you know or have a strong suspicion that there will not be foods that are included in your healthy way of eating, bring a dish that you'll enjoy and share it with the group. If it's not possible to bring your own dish and you don't want to eat the type of food that's being served at an event, eat a full meal before you walk out the door. Grab a drink, keep it in your hand at the party, and focus on socializing. Or, simply go to the party and try a few of the appetizers—in moderation. You'll be surprised how much this takes the pressure off you to be perfect. Choosing to fast in a nearly impossible situation where there are so many delicious treats can set you up for failure.

Sometimes, it is tough to remember that fasting is a choice. It's not something you have to do, and you certainly don't have to do it on a particular day. You have chosen to fast because you want to lose weight and improve your health. You should choose to fast on a particular day because the long-term benefits outweigh the short-term enjoyment of any particular food. You hold all the cards, and you have all the control.

How to Deal with Nosy Friends, Co-workers, and Partygoers

When I began fasting, I was surprised by the reactions and opinions of the people around me—at the office, at home, at parties, or just in everyday life when I chose not to eat. A lot of people didn't notice, but some did, and they had no problem sharing their views. These people had been used to me eating all the time, and they were confused when I passed on the candy jar, the birthday cupcakes, or the afternoon chips. At first, some *insisted* that I have a snack. Once they got used to my responses to snacks, though, they offered less, and not snacking got a bit easier for me.

If you're presented something to eat, try saying:

- "No, thanks, I'm full."
- "Thank you, but I'm not hungry."
- "It looks great, but I already ate."
- "I'm good for now."
- "It looks great, but I'm all set."
- "I may check it out later, but right now I don't feel like eating."

Try *not* to say:

- "I'm fasting."
- "I'm not snacking anymore."
- "I do not eat snacks."

If you do, you're opening yourself up for a big discussion (or confrontation!) and the well-intentioned snack bearers may feel that they need to defend their choice to nibble. Unless you're with your most trusted friends or loved ones, it's often easier to avoid the conversation entirely.

Adjust Your Priorities

Fasting is going to require that you alter some things yourself and how you spend your time. After all, if you changed nothing about your life, you would end up living in the exact same body you decided could use some improvement. If your social life consists of business lunches five days a week, happy hours with appetizers five days a week, weekend brunches with friends, Friday night dinners with your spouse, and Sunday all-day eating with your family, then yes, you are going to have to make major changes to your social life. You are going to have to adjust your priorities and choose fasting over a life that revolves around food. Luckily, if you're like me, you can look at your schedule and realize there are minor alterations you can make that will deliver big results.

As a businesswoman, I previously conducted most of my business meetings around food. I provided doughnuts, unhealthy lunches, and birthday cakes for every staff member, and I loved going out for lunches with current or prospective clients three or four times a week. When I started fasting, I cut back on a lot of these food-centered activities. Because I knew my staff still loved sharing treats and meals, I provided healthier alternatives, and instead of individual parties for each member of my staff, we celebrated once a month with cake. When I decided to skip lunch a few days a week, I began to change all my business lunches into coffees. No one seemed to care, and, in fact, few seemed to notice!

Know Your Limits

Many of us put ourselves in truly difficult situations to try to prove we are strong enough. This is a ridiculous and painful way to handle things. What if instead you treated yourself kindly and gave yourself the best possible chance of succeeding?

Now that I know I can fast successfully and have maintained my weight for the first time in my life, I know when fasting is the right choice for me. Personally, I do not fast for longer than seventeen hours at a time while on vacation or when my extended family is visiting because I know there will be delicious food around that I want to enjoy. I choose to do longer fasts when I am at home, usually just with my husband, and I can keep busy. These are the easiest times for me to be successful.

When do you have the best chances of success? This is when you should schedule your fasting. Fasting is not about saying no to yourself. Fasting is about saying yes to yourself on your own terms!

All these techniques may feel odd at first, but over time they become natural and you stop noticing that you have to do them. To truly change your mind and body, you must expect to make changes to your environment—but that's not a bad thing. It can be empowering to realize how much of your environment you can control. Give yourself the time you need to adopt the new behaviors and be patient with yourself. Soon, you *will* begin to enjoy your time at home, at work, at the grocery store, and at restaurants more than you would have ever believed possible because there will be fewer decisions and less stress. Notice your successes, even the small ones, and give yourself credit for making each change. These changes add up quickly, and suddenly, there is a new, happier you staring back from the mirror!

MEGAN RAMOS

My recommendation to my clients is always the same: fit fasting into your lifestyle, don't fit your lifestyle into your fasting schedule. Just as Eve said, fasting is flexible. That's why, with a little practice, you can find a balance between fasting consistently enough to reach your weight-loss goals and still maintain a healthy social life. Missing one fasting day is not a big deal, but you also can't miss them all.

Our lives are hectic, though, and it's hard to strike a balance between appointments, obligations, and fasting. I've found that sometimes overly organized people tend to be too inflexible about their fasting schedule. For example, here's a scenario I commonly hear:

CLIENT: "Megan, I'm feeling really sad. I missed my college reunion lunch on Wednesday."

MEGAN: "Why?"

CLIENT: "I fast for twenty-four hours every Monday, Wednesday, and Friday. You know that."

MEGAN: "Why couldn't you fast on Monday, Thursday, and Friday? Or Monday, Tuesday, and Friday?"

CLIENT: "Oh, I don't know why I didn't think of that. It didn't occur to me to be flexible with it."

On the flip side, some people can be too casual with their fasting schedule, letting it slide just because they associate a situation, a group of people, or a time of the year with feasting. For example, summertime is often problematic for fasters because people are so eager to socialize, barbecue, drink al fresco, and nibble at appetizers by the pool when the weather is nice. Here is another common conversation:

CLIENT: "I'm really frustrated that I haven't lost any weight. In fact, I think I've gained weight."

MEGAN: "What fasting regimen are you following?"

CLIENT: "I'm not really fasting. The weekends have been busy with barbecues and family outings."

MEGAN: "Well, what about Tuesdays?"

CLIENT: "I'm usually at work."

MEGAN: "Why can't you fast then? Just because all your summer weekends are busy with feasting doesn't mean that you can't find a day or two during the weekdays that you can fast."

CLIENT: "Oh, I hadn't thought about that."

The truth is that life is *always* busy. We often use the summer or weekends as an excuse, but we're highly social creatures all year long. Holidays are scattered every few months throughout the year as well. Being too rigid, too extreme, or too laid back is a recipe for failure. The people who get the best results from fasting are those who have flexibility and consistency and can plan and organize well. To start you off, here are a few examples of successful, balanced fasting regimens:

The Twenty-Four-Hour Fast, Three Times a Week

The twenty-four-hour fast doesn't need to be done on Mondays, Wednesdays, and Fridays. You can do two days back-to-back and a third day later in the week. You can even do all three days together, which works well if you have a busy long weekend ahead. Many successful people fast Monday, Tuesday, and Wednesday, and socialize guilt-free with friends and family for the rest of the week.

I don't recommend committing your fasting schedule to three particular days of the week, every week. Plan your weeks in advance, and if something unexpected pops up on a day you had decided to fast, then fast the next day in your schedule that isn't blocked off with social engagements.

The Thirty-Six- or Forty-Two-Hour Fast, Three Times a Week

The thirty-six- or forty-two-hour fast can be very successful, but it's tough to do it three times a week without disrupting your social life. A good strategy I've used myself is to fast for forty-two hours on Mondays and Wednesdays, or Tuesdays and Thursdays. Fridays and Saturdays are flex days, when fasting is an option depending on social engagements. I found it relatively easy to fast for forty-two hours consistently on Mondays and Wednesdays, but I struggled on Fridays and weekends, having tried and failed countless times. Finally, I decided I would just fast twenty-four hours on Fridays and be proud of it. Otherwise, I was being self-destructive on Fridays because a social situation would pop up about 80 percent of the time, so I'd either feel like I was missing out or blowing my fast.

Set a goal of fasting for twenty-four hours to forty-two hours, three times a week. Whatever you can do, be proud of it. Keep your attitude flexible and positive. Don't try to do too much because you'll only set yourself up for failure.

The Seventy-Two-Hour Fast Once a Week

Some people find it easier to extend a fast once they've started it rather than doing several smaller fasts. If you're one of these people, a seventy-two-hour fast may be the best option for you. Most people who practice this fasting schedule start the fast Sunday night after a weekend of good feasting, and they break their fast Wednesday night and eat for the rest of the week. This works great for people with very busy weekends with friends and family.

Three days of fasting sounds like a lot, but you're actually only skipping two weekday dinners, which is good for those whose dinner hour is important family time. It also works well if you're the primary cook at home since you have to plan in advance for only two meals during the week.

Planning for Holidays, Vacations, and House Guests

There will be times of the year when you just can't fast at all. These are occasions like holidays and reunions, or maybe you have overnight guests visiting from out of town. During these occasions, you should still avoid snacking. Why? Because you've worked hard to get your blood sugar level stable and reduce your insulin level; snacking will throw a wrench into that progress, and you'll start to feel hungry all the time again.

During holidays, my major piece of advice is to avoid grazing as much as possible. I know it's hard because food is everywhere, and so much of it comes in the form of sweets or junk food or greasy hors d'oeuvres—easy-to-eat foods that quickly become hard to keep track of once you start popping them into your mouth. But if you want to embrace and enjoy a feast, you need to focus on the one, full, satisfying meal. Eating less more often is the worst thing you can do for your progress.

Many of my clients have full-blown panic attacks before they go away on vacation. They've seen what a week of feasting over Christmas can do to them, and they're petrified that a week touring Tokyo is going to be twice as bad. Relax! The truth is that vacations are so much easier than holidays! Why? Because during Christmas, Thanksgiving, or many other major family or religious holidays, we're usually in a house, tempted to graze nonstop. We're presented with a massive spread of food from the moment we show up at our holiday celebration until the minute we leave. That's hours of eating! But when you're traveling, you're out and about, exploring, walking, visiting the local sights. You're only stopping to eat meals. That's it.

A client of mine recently returned from London, terrified to get on the scale.

"Megan, I drank beer and ate chips at every meal," he said.

Guess what? He lost seven pounds despite eating chips and drinking beer at every meal because he was only eating twice a day while he was away. He wasn't sitting on the couch in his hotel room every night snacking while watching Netflix. So try to go easy on yourself on vacation, and embrace the feasting. There will be times of the year when you can feast more, and others when you can fast more.

CHAPTER 21

Getting Back on Track

EVE MAYER

No one is perfect. And when you make any major lifestyle change—especially fasting and giving up the unhealthy foods you love—you're going to fail sometimes. It's okay, I promise. You can *always* get back on track.

One recent Saturday, I stumbled and fell off the wagon, too.

I was having a rough day. Years ago, my way of coping had been to binge on sugary and carb-filled foods, which took my mind away from my pain. I used food like most people used alcohol and drugs—sometimes in groups, but most of the time at home alone, in front of the television. Giving in to my constant hunger distracted me, and bingeing became an easy, comforting form of therapy.

Suddenly, that Saturday, my pain cascaded over me like a waterfall, becoming more intense the more it pounded into me. Panic enveloped me, and I became desperate for something, anything, to calm me down. Then it hit me: *I could binge.* Just this once—and never again—I could use food to stop my feelings. I was alone, so no one would know. I also had cash, so my husband wouldn't be able to see the credit card bill, with evidence of the bender I was about to head out on.

If you've never been addicted to a substance, then it's hard to describe the satisfaction—and, paradoxically, the desperation—that pills, liquor, or Twinkies can bring to an addict's life. But as I stepped on the accelerator and sped my car into the parking lot of a Nothing Bundt Cakes store, I felt that frantic urge, deeply. I slammed my door, then looked at my car and noticed that I'd parked it at an angle. I didn't care. I was crazed, ready to shove everything rational aside so I could get my sugar fix.

My heart sank when I went inside and saw six people in line. I didn't have that kind of time to wait—even to put that sweet Bundt cake with creamy icing into my mouth. Luckily, in the corner I noticed samples of red velvet Bundt cake with tiny dollops of white icing on them. I sprinted over, shoved one in my mouth, and swallowed without chewing. Then I turned my back to everyone and ate another sample, praying they hadn't seen me. Wiping crumbs from my mouth, I stepped back in line to wait my turn, noticed that there were still six people in front of me, and went back to the sample plate to devour two more tiny pieces of red velvet cake.

Then I scurried out of the store, hoping I'd become magically invisible.

You'd think I would have stopped there and headed to my car, but I didn't. I knew that in the same shopping center was an upscale grocery store that had food samples laid out around it. I walked through its doors and started bingeing on every sample they had—from yogurt to juice to beef jerky to grapes to my favorite: the fresh baked bread and cookies at the bakery.

I knew I was temporarily insane, and, yes, I was deeply ashamed of myself. But I was out of control, so I pushed those thoughts back down and reached for a piece of pumpkin bread.

Suddenly, I heard a voice near me. It was faint at first. Then it got closer.

"Eve! Hi, Eve, it's good to see you!"

I looked up and waved. To my horror, it was someone I knew. I was wearing no makeup and had my hair in a decidedly unattractive bun on top my head, and between those two things and my mouth stuffed full of bread, I should have looked like a monster. But I hadn't seen this friend in two years, and I was fifty pounds lighter than the last time we'd met up.

"Oh my gosh, you look fantastic!" my friend exclaimed.

"Thank you," I said, sincerely meaning it.

When I got home, I told my husband everything, from my pillaging of the samples of Bundt cake to my rampage through the grocery store. I cried as I told him how ashamed I was for cheating on a diet that had done so much for me, and I cried when I recalled how I'd been "caught" by someone I knew. As the tears flowed down my cheeks, though, he started laughing.

"Isn't it nice to be told how great you look when you have bread falling out of your mouth?" he asked.

It was true. It *was* pretty hilarious to receive a compliment at such a low point.

We chatted for a few more minutes after I dried my tears, and I started to take account of what had driven me to binge the way I had. Talking with my husband, I realized that I'd had some issues I'd avoided dealing with. I had concerns over my husband's new job, and my transition from a busy business owner to a consultant and full-time writer. I was losing my confidence and questioning my purpose, and those fears manifested themselves, allowing me to fall back into bad habits and let food control my life again. I felt like such a loser until I took a deep breath and realized: my binge had been just twenty minutes. That's it, twenty minutes of ridiculous, stupid behavior.

That's the lesson here. Even though you may have come *so* far—and still feel like you have a long way to go to be even healthier—you will always have minutes, days, or even weeks when you're not your best. You are only human, and that makes you imperfect. You are incredible, but you are also fallible. No matter how much you improve, how much work you put in, or how much you think you have your health all figured out, every once in a while, you are going to get knocked on your ass! And that's okay.

Hopefully you'll only have small problems, like my out-of-control binge fest. But other times, your problems will be huge. You may suffer from a disease, sustain a major injury, lose a job, end a relationship, or care for a sick family member. Life is a persistent, unforgiving roller coaster. Your job in the midst of this wild ride is to remember that you are worth the effort.

It is a cliché, but you are going to get off-track when you start fasting. I mess up, fail, and stumble. But now, instead of beating myself up for a lengthy amount of time, I simply start again. I have discovered that skinny people, who have been mysterious creatures to me for so long, pig out sometimes, too! Skinny, healthy people eat bad food and good food, but they follow a system of eating and/or fasting that works for them *most* of the time.

So, please, when you stumble and wonder if you'll ever get back on track, just tell yourself that it's okay to start again. Every big journey begins with just one step, so don't be afraid to get up, dust yourself off, and start over. You are worth it!

How to Stand Up After You Fall

1. Forgive yourself as soon as possible.
2. If you can learn to laugh at yourself, you will heal twice as fast.
3. Examine what is going on in your life and consider that you might not be getting something you need. If possible, try to give yourself what you need in a healthy way.
4. After you finish examining how you screwed up, recognize how you have succeeded. Give yourself a moment to be proud of your accomplishments.
5. Keep going. Try to do better than you did yesterday.

CHAPTER 22

Finding a Community

EVE MAYER

We've discussed the fact that fasting can be a polarizing topic. Because people tend to have strong reactions to the notion of skipping meals, sometimes it makes sense to keep your goals to yourself and not share the details of your new lifestyle widely.

That said, you also need and deserve support, and the more people who can help you on your journey, the better. In this chapter we'll discuss the best way to build a community.

Timing Is Everything

In my experience, it's better to share the fact that you're fasting *after* you have decided it's something that works for you. Your initial period of research, when you're educating yourself and deciding if you want to fast, is not the time to petition friends for their advice. Why? Because they don't live inside your body, and they haven't visited the depths of your mind. Like I've said before, too, they may be entirely uninformed about the health benefits of fasting.

If you're like me, taking a stab at losing weight or improving your health one more damn time puts you in an extremely vulnerable state. You might feel shame about your imperfect body, repeated failures, or supposed lack of willpower. When you're feeling exposed in this way, the opinions of others may carry more power than they should.

Hitting pause before announcing your new lifestyle to your friends or office mates allows you to exist in a vacuum of quiet where you can listen to your own opinions. For some of us, it takes great concentration to tune in to what our body and mind are trying to tell us, especially when we've been down on ourselves for so long. The choice of what you eat and when you eat is a singular, private, and personal one. Oftentimes, that decision can only happen when you listen to yourself, paying attention to your own needs and vision, for an extended period of time.

The best way to keep it private is to schedule fasting around your lifestyle. After all, canceling your appearance at a long-planned dinner party or avoiding your traditional Sunday brunch is going to raise some eyebrows *and* lead to a lot of questions you may not be ready to answer. Instead, plan to start your fast during a quiet, solitary, undisturbed time. If fasting works for you, you'll probably feel mentally prepared to talk about it with others sooner than you think.

• • •

Announcing Your Lifestyle

When you decide you're ready to form a community that understands and supports fasting, you might feel unsure about who's going to be receptive—or whether *anyone* will be. Yes, you can expect some strong reactions. Some people may express concern about the safety of what you're doing. Others will ask a million questions. And some people may get angry because your choice to fast makes them uncomfortable about their own choices. The inevitability of these reactions is why you need to know your stance first. If fasting has worked for you, stand firm and defend it. Don't let other people's questions or concern sway you. This is *your* choice and your health.

However, I encourage you not to be too defensive when you talk about fasting, for two reasons. One, when you're protective or apologetic about something, you often seem weak. Worse, you *feel* weak. That tentativeness does nothing to convince yourself—or others—that you've set out on the right path. Be sure of what you're doing and show it! Second, you might just be surprised about who's going to support you. I know I was!

A few weeks after I adopted intermittent fasting, I went to the office on a twenty-four-hour fast day, and there was free food in the conference room. The doughnuts that a kind co-worker brought in some mornings and the weekly birthday cakes my company provided contributed to a hefty portion of my weight gain, so I braced myself and worried, *What if someone asks me why I'm not eating?* While I could have kept my mouth shut, though, I decided I was firm in my stance about fasting and prepared to answer anything.

Not two minutes passed before a co-worker asked me if I

wanted a slice of the delicious-looking birthday cake. The moment had arrived. I was more than ready to broach the subject, so I answered.

"No, thank you. I've been fasting regularly for the last few weeks, and today I'm on a full-day fast." The room had been quiet, so I knew others had heard me. I braced myself for an onslaught of questions, funny looks, and insults that I was sure were about to come my way.

They didn't. Instead, I was astounded by the reaction in the room. My colleague smiled and shrugged like my fasting was no big deal, and a few other colleagues walked up to me and congratulated me. Some just ignored the issue entirely. I asked a few co-workers who approached me if they'd fasted, and some responded that, yes, they had for religious reasons when they were children. A few others were now regularly using intermittent fasting for weight loss, weight maintenance, or health.

A few people were curious about hearing more about my fasting, and others expressed the types of concerns that I had once worried about, too. They asked if my metabolism would slow down, did I feel faint or have a headache or feel hungry. These people were shocked to hear that I hadn't eaten for about eighteen hours and still felt great.

My group of co-workers were kind, supportive, and open, and I was lucky to have them. But, again, even smart, compassionate friends can be scared for your life when you start fasting. When I told my friend Jameelah that I hadn't eaten for four days, she begged me to stop fasting. As I continued fasting for a few days, she checked in with me and suggested repeatedly that maybe I could eat a little something or have some juice just to be safe. And when I went to visit my parents in Louisiana for the first time while fasting, these educated, open people who love me more than anything in the world were terribly concerned. I am their only child, after all!

That's why I recommend reading a few books, doing lots of research, following the blog and listening to the podcasts on FastingLane.com, and being firm in your convictions. You need to be equipped to deal with your most challenging critics as well as be prepared to be surprised by encouraging, understanding friends. But, most of all, know that if fasting is working for you, yours is the only opinion that matters!

Tips for Telling Others About Your Fasting

1. Educate yourself about all aspects of fasting—from its health benefits to the science behind it—before sharing your story with others. That way, you'll be prepared to answer unexpected questions.
2. Schedule your fasting around your life and not your life around fasting.
3. Wait until there is a solid, compelling reason that you need to share your fasting with others—if, for example, you won't be eating breakfast with your co-workers one morning or if you have to politely decline a piece of office birthday cake.
4. Expect emotional responses for or against your fasting.
5. Have at least one book and one online site you trust and can share with concerned friends or family members if they ask for more information.
6. Remember that what and how often you eat is ultimately *your* choice.

Build an Online Support Team

If you're anything like me, your entire life is contained on your smartphone. I keep my calendar and to-do list on it, I use it to check email, and every contact I've saved for the past ten years has been added to my phone rather than to an address book. My phone is often the last thing I see before I go to bed and the first thing I reach for when I wake up. It is my support system for every aspect of life.

Fasting is a significant lifestyle change, and many of us need help making it last. When it's going well, you're going to feel like you're on top of the world, and when it isn't, you're going to feel frustrated and need the support of people you trust who are available when you need them. Real-life friends and family should always be your first choice, but, eventually, they may get sick of you talking about what you ate or didn't eat, and how you feel. That's why I recommend finding a community online.

I first learned about fasting through a friend who suggested I read *The Obesity Code* by Dr. Fung. I then looked Dr. Fung up online and read about him, watched his videos, listened to his podcasts, scoured his website, and learned of his partner, Megan Ramos. I also joined his online fasting support groups, and they led me to some of my favorite sites, like DietDoctor.com, and Facebook groups specific to low-carb, keto, and fasting. Through Facebook groups, I met people like me who were new to fasting and had a million questions. I began interacting with people from all over the world who had been fasting for years and had so many answers. Today, I believe that online resources are a huge reason I'm experiencing lasting success with my weight loss for the first time in my life. In fact, they've been so important that I started my own site and podcast at www.FastingLane.com, which collects the low-carb and fasting doctors, pros, and successful everyday people whom I admire and have learned from. I'd love for you to join us there and perhaps to share your own story someday.

To develop the best online support system, start with medical professionals. Find doctors and researchers who explain the science of fasting and food in a way that you understand. Make sure their words resonate with you and test their theories to prove them to yourself. I don't expect that you will believe every word any person will say, but I believe you will find a small group of professionals who will help you find the best guidelines for improving your health.

Next, look for people on a social media platform you use daily who are online personalities living the type of lifestyle you want to live. For instance, if you use Twitter and are doing extended fasting, follow someone who does the same and tweets about it. Read this person's tweets, and if you get really bold, interact with them.

The easiest place you can do this is Instagram. There are loads of people who post specifically about any type of eating or fasting you can imagine. If you decide to eat low-carb, for example, you can search this term and find all the people you could ever want discussing this topic on a daily basis. Find a few whose recipes look good to you and whose style of posting content you find entertaining and informative. Some people will also offer tips, give encouragement, tell stories, and offer discounts. Others will just post "before" and "after" pictures with measurements and share details of their fasting regimen. Find people you look up to and can learn from. You get to pick your inspiration, and that can mean one person or several who each give you a little bit of what you need.

When you want to interact with a group of people who have

similar goals as you, I recommend that you turn first to either a paid site run by a professional group you trust or a Facebook group. Facebook groups have been surprisingly helpful to me. Bear in mind that a Facebook page is different from a Facebook group. The purpose of a Facebook page is usually for a brand or organization to get their message out to the people who choose to like the page, while the purpose of a Facebook group is to build a community for a group of people with something in common. The organization that owns a Facebook group will sometimes post, but the group is intended for like-minded people to help one another.

I find it very comforting to hear how others feel when they fast. I like to know that at hour 36 of a forty-eight-hour fast many people feel very hungry and are tempted to give up. It is one thing to recognize that common experience with a single person, but when I see it repeated from all types of people at different times in a Facebook group, I feel validated. Reading strangers' stories online makes me feel less like a freak because we all have so much in common.

When you begin to fast, it can be very helpful to have people online who are fasting along with you. You will very often experience the same frustration, hunger, success, and joy at the same times. It is comforting when you decide to do your first week of three twenty-four-hour fasts and you get to do it with a group of people you have never met. Some of your questions and fears are easier to tackle with online strangers. I find it very encouraging to answer questions online because I'm able to share the knowledge I've collected, and I know how that person feels because I was once there. It is a reminder of how far I have come. Watching people post their "before" and "after" pictures can also be a huge source of inspiration. After all, if they can achieve their goals, you can, too.

Finally, an online support system will help you celebrate the little NSVs (non-scale victories) that your real-life friends might grow weary of. These include victories like needing only half of your diabetes medication daily, or being able to take a selfie that you are proud of. An online support system is there twenty-four hours a day, ready to answer your questions, hear your frustrations, and celebrate your wins while you help them do the same. You deserve all types of friends—even the ones you will never meet in person but who will cheer your victories every step of the way!

Three Tips for Building Your Online Support Squad

1. Choose a few doctors and researchers to follow on fasting and eating who resonate with you and sign up for their online newsletters or join their group.
2. On your favorite social media platforms, follow two chefs or recipe developers who specialize in your type of fasting and eating.
3. Join an online group or two on Facebook where you can ask questions and exchange support for people going through the same experiences as you.

CHAPTER 23

Living Your
New Life

EVE MAYER

It wasn't easy for me to adjust to my new fasting lifestyle. Why? Because I don't smoke, I rarely drink, and I don't bungee jump. When I go on vacation I want to relax, and I want to EAT! What was I going to do now that nonstop snacking and gorging were no longer on the table?

All I had to do was open my eyes and look around to discover a new source of joy—one that made me truly, deeply happy in my own skin.

My first vacation was a three-week road trip in a camper with my husband. Now, this was no ordinary van. With a foldout bed, a hidden toilet, a fridge, a microwave, counters, and, most important, a coffee maker, it defined the word *glamping*. Solar panels on top allowed us to enjoy full power and internet, but that still wasn't enough for me. Road trips used to mean Cheetos, chocolate, salt and vinegar chips, and endless amounts of soda and coffee, and my inability to stock the mini fridge made me feel nervous, disappointed, and, quite honestly, not nearly as excited about vacation as I normally am.

I learned on day two that a surprising amount of the country does not have great cell phone service or access for my hotspot, so I talked to my husband and sat with my thoughts, which were mostly about food. Still, I forced myself to focus on other things, so I looked out the window at the sky. It was *beautiful*. It was purple in the morning and pink in the afternoon, and as we drove through Wyoming to Montana, more and more birds filled it. As we crossed into Canada, gorgeous mountains extended into it, disappearing into the clouds when we arrived in Banff.

The times we ventured out of the camper were even more amazing. In Yellowstone National Park, I gasped at the sight of buffalo and giggled when people got sprayed by a nearby geyser at Old Faithful. I stared at snowcapped mountains from the back porch of a library while my puppy romped with a new doggy friend through the tall grass. I sipped coffee with no sweetener in crisp summer mountain air, in awe at the sight of Crazy Horse and Mount Rushmore. I slept late, kissed my husband, went on long walks, looked at the fish in the clear water of the river, listened to a cellist play in the library, and watched movies in the van. I played Frisbee with the puppy and bought a new backpack. And, slowly but steadily, I thought less about food and more about what was around me.

It was like I was coming back to life.

I had never really appreciated nature. It's for fit, healthy people, after all, and it's hard to appreciate the view when you can barely breathe because your body is too heavy to carry you up a tiny hill. It's hard to think about things other than food when your hormones are conditioned to tell you to eat every damn minute of the day. It is hard to appreciate your dog, your husband, your child, your life, your anything when every moment your mind is screaming at you to EAT! But I started fasting, and I found out that I could do so much better. Life was different from what I thought it was, and it was during that vacation that I started to feel something new.

It was freedom.

Fasting has helped me to develop an appreciation for many things I'd been taking for granted. I still love to eat, but, for me, there's so much more to life than that. Fasting is going to give you more free time and money to use in whatever way you wish. But, above and beyond that, it's going to give you the freedom to get out there and enjoy life.

Let's celebrate!

How to Celebrate

In the past, when I achieved something in life, there was only one way to celebrate. Cake! And champagne. And steak. And a loaded baked potato. You get the idea. Once you change to eating healthily and fasting, celebrating every little success by gorging on food no longer works.

You need new ways to celebrate!

You are going to have a lot of celebrating to do. You need to honor your big goals, but you can also celebrate small achievements along the way. Maybe your big goal is to lose eighty pounds, but your interim goal is to bench-press eighty pounds. The celebration should be commensurate with the achievement. Reward yourself with something small when it takes little effort to achieve it. For example, you might get the remote control to yourself for a night when you complete your first twenty-four-hour fast. Celebrate with something big, though, when it takes time and persistence to get there. For example, you might mark losing twenty pounds with tickets to see your favorite band play live.

I'll be honest; I still celebrate with food, but I've found a better, healthier way to do it. It seems pretty strange to reward weight loss by eating, but there's something kind of cool about enjoying a big medium-rare rib-eye steak and a wedge salad at a fancy restaurant after a long fast.

Celebrating is a big deal because it reminds you of where you came from and what you achieved. It reinforces your confidence and your commitment. I recommend honoring your achievements with someone who supported you in your goals as a way of thanking them. Set aside some of the money you save from fasting and put it in a bank account or an actual piggy bank for your celebration fund. It doesn't matter if you can afford two bucks a week or $200 a week. You can watch the savings toward your eminent success grow as you get healthier.

How you will celebrate depends on what you enjoy doing. But regardless of whether you buy a new outfit, take a long bath, or buy yourself flowers, do *something* to commemorate all that you've achieved. You deserve it!

AFTERWORD:
IS BARIATRIC SURGERY
FOR YOU?

EVE MAYER

I've had bariatric surgery not once, not twice, but *three times*. Yes, you read that right, *three times*. Before I discovered fasting, I failed to keep weight off through diets, so I did some research and learned that most studies show that people lose weight after bariatric surgery. I visited multiple doctors, got the support of my family and friends, and decided to go for it. In 2004, I underwent my first procedure, then waited patiently to see if the pounds would fly off.

They did—at first. Going under the knife also allowed me to go off medications and reduce the side effects of polycystic ovary syndrome, which helped me become pregnant with my daughter. These were the upsides—and they were *huge*. But the bad news was that, all three times I had bariatric surgery, I eventually gained back some or all of the weight. The relentless hunger I'd felt my whole life, which had faded away right after surgery, returned as well. In other words, the surgery didn't fix my "broken" body. That would happen years later, when I discovered fasting.

If you are considering bariatric surgery, I urge you to not only do the necessary research but also speak with others who have been through it. I hope my story will show you its positive and negative aspects, and help you decide whether it's the right choice for you.

My Story

My first bariatric surgery was a lap-band (short for "laparoscopic-band") procedure. In lap-band, doctors insert a band into your body around the upper part of your stomach. The band can be adjusted to be tighter or looser by adding saline through a needle and into a port that's placed just under your skin on your stomach. When working properly, a lap band should make you feel fuller more quickly and for longer. The surgery is minimally invasive and is usually done with a few small incisions on the abdomen. It can also be reversed if there is ever an issue.

During the first few months after the procedure, I was ecstatic. My recovery time wasn't more than a couple of weeks, and the pain was bearable. For three months after the surgery, I was hungry less often, and I went from about 300 pounds down to around 225.

However, in the fourth month after my first lap-band surgery the hunger returned with a vengeance. My weight loss slowed and, over the next few months, stopped completely. Then the weight began to creep back up. Within four years, it had stabilized at around 230. My goal had been to get to 185, so this was a failure in my book.

There were side effects to lap-band surgery. If I ate rapidly—especially fibrous or dry food like chicken breasts or broccoli—or consumed too much, I threw up. If the lap band was adjusted too tightly, I also vomited. With the lap band, nighttime was brutal because if I ate within three hours of going to bed, my dinner would come right back up. As I lost or gained weight, the band needed to be adjusted since my internal organs also contracted or expanded. I found it very difficult to find the sweet spot with the lap band, so, over time, I got fewer adjustments and did what many people do. I worked around the lap band, not with it.

Since I was no longer losing weight and now gaining, I began speaking with the surgeons where I worked. They suggested I get my lap band taken out and have the new, improved version put in. So I did, and my second bariatric surgery in 2007 removed my original lap band and replaced it. As before, I thought this would be a permanent solution to my hunger, and once again, it worked like magic! My weight went down to about 185, and I was *ecstatic*.

It didn't last. I stayed at 185 for exactly two days, and then I began the journey back up again to 225. My diet with the lap band was also less healthy than it had been pre–lap band. Why? Because I was motivated to try to eat, feel full, and not get sick. I ate hamburgers with extra mayo and greasy fried meats with as much sauce as possible. I knew I was making bad choices, but I missed eating, and I was *hungry*.

That's right. After about three months with the new lap band, I was just as hungry as I had been before. The only difference was that now my stomach was small and filled up quickly. This meant that I ate all day long.

Six years went by, and I continued to struggle with steadily increasing weight. I was a prisoner to my constant obsession with food, so I decided to research bariatric surgery for a third time. I considered having a gastric bypass, a surgery that divides your stomach and makes the portion in which food is digested smaller. It's a complicated, serious procedure, so I opted instead to have my lap band removed and get a gastric sleeve, an easier surgery in which the stomach is made narrower, like a cylinder. A gastric sleeve is typically done with small incisions and does not require more than a couple of weeks of downtime. It is not adjustable, and I liked the idea of not having the issues I'd had with the lap band. Unlike my lap band, the gastric sleeve is not reversible, but I was determined to solve my hunger issues once and for all.

All in all, I had a good experience getting my gastric sleeve. The doctor was proficient and kind, and I did not experience a lot of pain. I was back to full speed in a couple of weeks. For the first few months I felt much less hungry, and my weight dropped to under 200. But after about three months, the constant hunger was back. Sure, I was happy to find that I could consume more fibrous and healthier foods than I had before, but I was crushed that I still felt famished.

As of this writing, I've had a gastric sleeve for about seven years. The gastric sleeve was a much more effective surgery for me compared to the lap band, but the truth is that three bariatric surgeries never fixed my body or my mind. If I could get in a time machine and go back fifteen years, I would spare myself the risk, the money, and the pain of those surgeries. Fasting was the ultimate solution for me, and if you're considering bariatric surgery, I urge you to try fasting first.

JASON FUNG

It is often said that bariatric surgery is the only scientifically proven method of long-term weight loss, but I have found that it's rife with problems.

The first is that it's not usually successful in the long term. The Cleveland Clinic has stated that gastric bypass surgery carries a 10 to 20 percent risk that you will require a follow-up operation to correct complications, that about a third of patients will get gallstones, and that almost 30 percent will experience nutritional deficiencies. With lap band, they claim that most of their patients experience at least one serious side effect, including bowel obstruction, hair loss, bleeding, and clots.

My own clinical experience with bariatric surgery tells much the same story. Almost all my clients who have undergone lap banding have already had it removed. One of my clients who had gastric bypass surgery required repeated procedures, and because her stomach kept scarring, she had persistent nausea and vomiting. So far, I have not met a single person who maintained a clinically significant weight loss using surgical techniques.

The numbers back this up because the success rate over the long term is low, and even for the most powerful gastric bypass surgery, there is a significant risk of weight regain that increases over time. For individuals ten years out from bariatric surgery, the failure rate is up to 35 percent.

The problem is not that we need better surgical techniques, but that we need to change the entire culture of weight loss. The medical profession is so steeped in the "Calories In, Calories Out" philosophy of weight loss that, when it fails (which it inevitably does), doctors resort to drastic and draconian measures of mutilating a normal body for the purposes of weight loss. To me, this represents an admission of defeat, and it's simply not good medicine. Doctors are supposed to prevent their patients from having surgery, not endorse it, right?

Luckily, the poor performance rate of bariatric surgery seems to be catching up with the numbers. In 1999, bariatric surgery was virtually unheard of, with only 6.9 procedures per 100,000 people. In five short years, that rate increased more than tenfold, to a rate of 71.06 per 100,000.

Then a funny thing happened.

Growth in bariatric surgery completely flatlined after 2003. The zenith was reached in 2009 at a rate of 71.26 surgeries per 100,000 people, then started a slow, inexorable decline. Why? The reason was clear enough. Surgery for obesity just didn't work very well, and the people who'd had it were clearly warning their friends to avoid it. The word on the street was out. Bariatric surgery wasn't good. Bariatric surgery rates have been slowly declining since then.

I often think of a client, Pamela, who weighed 335 pounds when we first met. Pamela wasn't experiencing obesity-related conditions like hypertension, type 2 diabetes, or fatty liver. But she'd developed asthma, joint pain, depression, low energy, and irregular periods. Vowing to lose weight, she set herself a goal of losing 135 pounds. Her doctor said that if her progress had stalled or stopped in six months, they should discuss bariatric surgery.

She told herself that she'd never go that route.

Instead, she dropped refined sugars from her diet, drastically reduced her carb intake, focused on eating healthy fats, and began intermittent fasting. One year and twenty-six days after changing her lifestyle and diet, she reached her goal. At 200 pounds, she's lost twenty-three inches off her waist. Today, Pamela fasts for forty-two hours three times a week and often extends those fasts longer. She has loads of energy, and she's decided to keep losing weight until she hits 150 pounds. Best of all, she's done all that without going under the knife!

The success of people like Pamela and the failure of bariatric surgery for thousands of people are among the many reasons why it's time to consider fasting for its proven long-term success. It's high time for fasting to enter the mainstream, be endorsed by doctors and the medical establishment, and gain recognition as the simplest way to maintain weight loss and great health. If you're overweight or suffering from chronic health complaints, I urge you to try it, too.

ACKNOWLEDGMENTS

EVE MAYER

My deepest thanks go to my dear friend, Dr. Suzanne Slonim, who first introduced me to Dr. Fung's work. Thank you to Dr. Jason Fung and Megan Ramos, my gurus of fasting, who were so kind to pursue this book in partnership with me.

To my husband, Levi Sauerbrei, my partner in life and in fasting, who held me all the days I cried because I was angry that I hadn't discovered fasting sooner: Levi, thank you for supporting my dream of helping others with FastingLane.com. Thank you to my friend and team member, Bridgette Hardy of FastingLane.com, who dived into fasting and surpassed my own skills. From the bottom of my heart, thank you to my daughter, Luna, who at only thirteen years old has used my experience to change her own eating for better health. Luna is forever grateful I no longer force her to eat breakfast when she isn't hungry. To my parents, Guy and Regina Mayer, who were scared for me when I said I wasn't going to eat for ten days but then stayed open: thank you, thank you. My parents watched my experience, asked questions, offered support, and made changes to their own way of eating for better health.

Thank you to our agent, Rick Broadhead, for his tireless work on this book. And thank you to the talented Sarah Durand for making our book even better!

Thank you to HarperCollins for supporting this message and delivering it into the hands of so many people who can have longer and more fulfilling lives because of it.

MEGAN RAMOS

This book would not be possible without the hard work and dedication of Dr. Jason Fung. Not only was he fearless in combating the current standard of care to fight for the best medical practices for patients, but he also saved my life. Having been diagnosed with type 2 diabetes at a very young age, I was on a fast track to an early demise. Today I'm healthy and strong. Even when people said fasting was crazy, Jason kept talking about it. He spent countless hours researching to provide the most sound, well-referenced scientific explanations possible. He lifted the sugary fog that was impeding my vision. And because of this dedication, I get to spend my days helping people take control of their health and overcome chronic disease that would otherwise kill them.

After working with thousands of patients from all over the world, I've learned one thing: people need to know they're not alone in this. Eve Mayer's selflessness and transparency in sharing her incredible story in this book fill that void. Her story is something I wish every single person I meet in our clinic could hear. She hasn't just amazed me with her incredible story, passion, and southern charm through this process; she's also become a mentor and a best friend.

I'd like to thank our team: Rick Broadhead, Sarah Durand, and our incredible editor at Harper Wave, Julie Will, for making this book and getting it out there for the thousands of people who need to hear Eve's story. I hope one day they realize the number of lives they have helped us save.

To the thousands of people whom I have worked with through Intensive Dietary Management and The Fasting Method, thank you for trusting me to help you. Thank you for letting me learn so that I can go on to help so many more people. Thank you for making every day never feel like a workday.

Thank you to the readers of this book, who have yet to give up on themselves. You are my heroes, and you deserve to feel good in your body!

Last, but not least, I'd like to thank my incredible husband, Angel, for joining me on this wild ride. I can't count the number of times he's made dinner or done errands alone so that I can work on this book. He's my rock.

JASON FUNG

I'd like to extend my deepest gratitude to my co-authors, Eve Mayer and Megan Ramos, for making this book a reality.

At HarperCollins, special thanks to my editor, Julie Will, with the able assistance of Haley Swanson and Emma Kupor. Thank you as well to Andrea Guinn, Brian Perrin, Yelena Nesbit, David Koral, and Bonni Leon-Berman.

FASTING GLOSSARY

5:2 DIET: Eating five days a week and fasting the remaining two days.

16/8: Sixteen hours of fasting with an eight-hour eating window.

20/4: Twenty hours of fasting with a four-hour eating window.

24: Fasting for twenty-four hours.

ACETIC ACID: The main component of vinegar, outside of water. It gives vinegar its pungent smell and sour taste. Vinegar can be used to suppress appetite and slow the digestion of starchy and refined carbohydrates, which lessens the spike of blood sugar as the carbohydrates are digested. Vinegar deactivates an enzyme called amylase, forcing your pancreas to produce the amylase and slowing the digestion in your small intestines.

ALTERNATE-DAY FASTING (ADF): Alternate-day fasting involves fasting one day, eating the next, and repeating the pattern.

AMINO ACIDS: Organic compounds that make up proteins. When broken up during digestion, the liver reassembles the amino acids to make new cellular proteins such as blood cells, bone, muscle, connective tissue, skin, and more.

APOPTOSIS: A form of programmed cell death that occurs in multicellular organisms.

APPLE CIDER VINEGAR (ACV): Apple cider vinegar has a long history as a home remedy, used to treat multiple ailments. It is thought to help regulate blood sugar and improve your digestion.

AUTOPHAGY: This is the body's mechanism of getting rid of all of the broken-down, old cell machinery (organelles, proteins, and cell membranes) when there's no longer enough energy to sustain them. It is a regulated, orderly process to degrade and recycle cellular components. It starts to occur after approximately twenty-four hours of fasting.

BLOOD GLUCOSE (BG): Also known as blood sugar, blood glucose is sugar absorbed into cells via insulin to feed all major tissues and the brain. The only major tissue not fed by blood glucose is the liver. Excess blood glucose is stored in the liver and turned into fat.

BONE BROTH: A broth made from simmering the meat bones of animals with vegetables, herbs, and spices for several hours. It provides nutrients during fasting.

BULLETPROOF COFFEE (BPC): Coffee mixed with butter and MCT oil. Intended to add additional fats to coffee for nourishment.

CALORIES IN, CALORIES OUT (CICO): The commonly held belief that "calories in" minus "calories out" equals the amount of fat stored or fat lost (deficit).

CHIA SEEDS: Good source of additional fiber that cannot be broken down into refined sugar; ingesting it also helps a person feel fuller longer.

DRY FASTING: Going without food or drink for an extended period. This type of fast combines fasting with light dehydration and is not recommended.

ELECTROLYTES: Certain minerals in the bloodstream that include sodium, chloride, potassium, calcium, magnesium, and phosphorus. During fasting, your electrolyte levels may become low.

ERYTHRITOL: Made from fermented corn or cornstarch, erythritol is a sugar alcohol that occurs naturally in small quantities in fruits and fungi like grapes, melons, and mushrooms. It is only partially absorbed and digested by the intestinal tract, which can cause gastrointestinal discomfort in some people. It is known as a low-carb approved sugar substitute.

EXTENDED FASTING (EF): Extended fasting consists of fasts longer than seventy-two hours (three days).

FAT ADAPTED: When your body shifts into burning fat for energy instead of glucose.

FEASTING: The opposite of fasting; days when you eat food.

GHRELIN: The hunger hormone that turns on appetite.

GLUCOSE TOLERANCE TEST (GTT): Identifies how someone's body handles glucose (sugar) after a meal. The test is an oral test conducted after at least eight hours of fasting.

GLYCOGENESIS: The creation of glycogen in the liver. Insulin is the main stimulus of the creation process.

HANGRY: Hungry to the point of becoming angry.

HC/HWC: Heavy cream/heavy whipping cream.

HIGH-DENSITY LIPOPROTEIN (HDL): Cholesterol level measurement often called the "good cholesterol."

HYPOTHYROIDISM: Decreased metabolism as a result of thyroid hormone deficiency.

INFLAMMATION: A localized physical condition in which part of the body becomes reddened, swollen, hot, and often painful, especially as a reaction to injury or infection.

INSULIN RESISTANCE: When cells are no longer responsive to insulin and normal amounts of insulin are not able to move glucose into cells, resulting in the buildup of glucose in cells. To compensate, the body has to produce more insulin, leading to a constant high level of insulin, which blocks fat burning. If your fasting blood sugar is 5.7 (103) and your insulin is high, too, over 12 μU/mL, you are insulin resistant and on your way to type 2 diabetes. If your blood sugar is 5.7 but your fasting insulin is under 9 μU/mL, you are insulin sensitive and likely in glucose refusal mode from a low-carb diet.

INTERMITTENT FASTING (IF): Time-restricted eating cycled with longer periods of voluntarily not consuming food. IF focuses on *when* you eat, not *what* you eat.

KETO/KETOGENIC: Eating a diet consisting of 75 percent fat, 20 percent protein, and 5 percent carbohydrates, resulting in the body entering a state of ketosis. A ketogenic diet is a very-low-carb diet and may contain 20 grams of carbohydrates or less. A person's body switches its fuel supply to burning fat. Insulin levels drop significantly and fat burning increases dramatically.

KETONE: An alternative fuel source produced in the liver when the body is burning fat. Ketones fuel the brain when glucose is low and are produced when low amounts of carbohydrates, moderate amounts of protein, and high amounts of fats are eaten.

KETOSIS: A metabolic state entered when the body produces ketones.

LEPTIN: The hormone that travels to the brain to signal we are full. Upon reaching the brain, leptin decreases appetite, stopping eating and lowering insulin.

LOW CARB HIGH FAT (LCHF): A low-carb, high-fat diet.

MACRONUTRIENT (MACROS): Proteins, fats, and carbohydrates, the three components of human diets.

MCT OIL: The medium-chain triglycerides of coconut oil and palm kernel oil extracted from their original state of fat.

METABOLIC SYNDROME: Insulin resistance, also called prediabetes, resulting in a cluster of co-occurring symptoms, including increased blood pressure, high blood sugar, excess body fat around the waist, and abnormal cholesterol or triglyceride levels.

METFORMIN: Blood-sugar-lowering medication commonly used in type 2 diabetes management.

MONK FRUIT SWEETENER: A relatively new sugar substitute derived from a round, green fruit grown in Southeast Asia. It has noncaloric compounds called mogrosides that provide its intense sweetness. It is considered an LCHF-approved sugar substitute.

NET CARBS: Total carbohydrates minus grams of fiber and sugar alcohols.

NON-SCALE VICTORY (NSV): A weight-loss success that is not a number on a scale (e.g., a smaller clothing size or being able to run one mile).

ONE MEAL A DAY (OMAD): Fasting where a person eats only one meal a day.

PERIODIC FASTING: Another name for intermittent fasting.

POLYCYSTIC OVARY SYNDROME (PCOS): PCOS is the most common reproductive disorder in the world and is a hormonal disorder causing enlarged ovaries with small cysts on the outer edges. It affects an estimated 8 to 20 percent of women of reproductive age.

STANDARD AMERICAN DIET (SAD): Based on the food pyramid and several studies conducted in the 1970s and early 1980s, the diet references how most Americans eat. It recommends a minimum of three meals a day, with most calories and elements being refined carbohydrates. The SAD has led to epidemic levels of obesity, hypertension, heart disease, and diabetes.

STEVIA: A low-calorie sweetener made from the leaves of the stevia plant. In the United States, the active sweet compounds are extracted and processed into liquid or powder.

STRICT KETO: Following a ketogenic diet fully to include the elimination of sugar, grains, starchy vegetables, and processed foods while keeping daily carbohydrates at or under 20 grams and eating natural protein sources.

TIME-RESTRICTED FEEDING: Intentionally cycling between eating and fasting states, focusing on when a person eats, not what they eat. This is also called intermittent fasting.

TOTAL DAILY ENERGY EXPENDITURE (TDEE): Total number of calories expended every day.

TYPE 2 DIABETES: A condition in which insulin levels are high and the body has become insulin resistant. Blood sugar is elevated and doesn't allow insulin to do its job. It differs from type 1 diabetes in that the body can still create insulin.

WATER FAST: A fasting cycle supported by water only.

WAY OF EATING (WOE): Describes what, when, and how someone consumes food.

SOURCES

Chapter 1: The Science of Fasting

Calle, E. E., C. Rodriguez, K. Walker-Thurmond, M. J. Thun. "Overweight, Obesity, and Mortality from Cancer in a Prospectively Studied Cohort of U.S. Adults." *New England Journal of Medicine* 348, no. 17 (April 24, 2003), 1625–38.

Green, M. W., N. A. Elliman, P. J. Rogers. "Lack of Effect of Short-Term Fasting on Cognitive Function." *Journal of Psychiatric Research* 29, no. 3 (May–June 1995), 245–53.

Lieberman, H. R., C. M. Caruso, P. J. Niro, G. E. Adam, M. D. Kellogg, B. C. Nindl, F. M. Kramer. "A Double-Blind, Placebo-Controlled Test of 2 d of Calorie Deprivation: Effects on Cognition, Activity, Sleep, and Interstitial Glucose Concentrations." *American Journal of Clinical Nutrition* 88, no. 3 (September 2008), 667–76.

Nassour, J., R. Radford, A. Correia, et al. "Autophagic Cell Death Restricts Chromosomal Instability during Replicative Crisis." *Nature* 565 (2019), 659–63. doi:10.1038/s41586-019-0885-0.

Singh, R., D. Lakhanpal, S. Kumar, S. Sharma, H. Kataria, M. Kaur, G. Kaur. "Late-Onset Intermittent Fasting Dietary Restriction as a Potential Intervention to Retard Age-Associated Brain Function Impairments in Male Rats." *Age* 34, no. 4 (August 2012), 917–33. doi:10.1007/s11357-011-9289-2.

Chapter 2: Beyond Science

Judge, T. A., D. M. Cable. "When It Comes to Pay, Do the Thin Win?: The Effect of Weight on Pay for Men and Women." *Journal of Applied Psychology* (2010).

Roehlin, P. V., et al. "Weight Discrimination and the Glass Ceiling Effect among Top US CEOs." *Equal Opportunities Int.* 28, no. 2 (2009), 179–96.

Roehling, M. V. "Weight-Based Discrimination in Employment: Psychological and Legal Aspects." *Personnel Psychology* 52 (1999), 969–1016.

Watkins, Ellen, and Lucy Serpell. "The Psychological Effects of Short-Term Fasting in Healthy Women." *Frontiers in Nutrition* 3, no. 27 (2016).

Chapter 3: Hormones and the Hunger Bully

Espelund, U., et al. "Fasting Unmasks a Strong Inverse Association between Ghrelin and Cortisol in Serum: Studies in Obese and Normal-Weight Subjects." *Journal of Clinical Endocrinology Metabolism* 90, no. 2 (February 2005), 741–46.

Natalucci, G., et al. "Spontaneous 24-h Ghrelin Secretion Pattern in Fasting Subjects: Maintenance of a Meal-Related Pattern." *European Journal of Endocrinology* 152, no. 6 (June 2005), 845–50.

Chapter 5: A Path to Healthier Eating

JACC Study Group. "Dietary Intake of Saturated Fatty Acids and Mortality from Cardiovascular Disease in Japanese: The Japan Collaborative Cohort Study for Evaluation of Cancer Risk." *American Journal of Clinical Nutrition* 92, no. 4 (October 2010), 759–65, https://doi.org/10.3945/ajcn.2009.29146.

Mozaffarian, D., E. B. Rimm, D. M. Herrington. "Dietary Fats, Carbohydrate, and Progression of Coronary Atherosclerosis in Postmenopausal Women." *American Journal of Clinical Nutrition* 80, no. 5 (November 2004), 1175–84.

Schulte, E. M., N. M. Avena, A. N. Gearhardt. "Which Foods May Be Addictive? The Roles of Processing, Fat Content, and Glycemic Load." *PLoS ONE* 10, no. 2 (2015). https://doi.org/10.1371/journal.pone.011795.

Chapter 8: Get Your House in Shape and Your Family on Board

Ho, K. Y., J. D. Veldhuis, M. L. Johnson, R. Furlanetto, W. S. Evans, K. G. Alberti, M. O. Thorner. "Fasting Enhances Growth Hormone Secretion and Amplifies the Complex Rhythms of Growth Hormone Secretion in Man." *Journal of Clinical Investigation* 81, no. 4 (April 1988), 968–75.

Afterword

Christou, N. V., et al. "Weight Gain After Short-and Long-Limb Gastric Bypass in Patients Followed for Longer Than 10 Years." *Annals of Surgery* 244, no. 5 (November 2006), 734–40.

Johnson, E. E., et al. "Bariatric Surgery Implementation Trends in the USA from 2002 to 2012." *Implementation Science* 11, no. 21 (2016).

Smoot, T. M., et al. "Gastric Bypass Surgery in the United States, 1998–2002." *American Journal of Public Health* 96, no.7 (July 2006), 1187–89.

INDEX

ABOUT THE AUTHORS

JASON FUNG, MD, was born in 1973 and trained in Los Angeles and Toronto as a kidney specialist. He founded The Fasting Method (www.TheFastingMethod.com) to provide evidence-based advice for weight loss and managing blood sugars, focusing on low-carbohydrate diets and intermittent fasting. It had become obvious that conventional medical treatments were failing patients. Although many of today's chronic medical issues are related to diet and obesity, treatments are focused on medications and surgeries. If you don't deal with the root cause, the problems never improve. A dietary problem requires a dietary solution. Dr. Fung is the author of *The Obesity Code*, *The Complete Guide to Fasting*, and *The Diabetes Code*. He is also the scientific editor of the *Journal of Insulin Resistance* and the managing director of the nonprofit organization Public Health Collaboration (Canada), an international group dedicated to promoting sound nutritional information.

EVE MAYER is an author, speaker, humorist, and entrepreneur. She lives in Carrollton, Texas, with her husband, Levi, her daughter, Luna, and her mutt, Holly. Eve is the author of *Social Media for the CEO*, *The Social Media Business Equation*, and *Get It Girl Guide*. Eve is a consultant to companies at the intersection of marketing and company culture, enabling them to build better core values, market honestly, and embrace diversity. For her entire adult life, Eve struggled to overcome obesity and poor health. Despite experiencing success in most aspects of her life, her failure to take control of her own health and body brought feelings of deep shame. After twenty-four years of dedicating herself to diets, therapies, punishing exercise regimens, and three bariatric surgeries, Eve found lasting wellness through a combination of intermittent fasting and low-carb eating. Eve once fasted publicly for ten days straight under the coaching of Dr. Jason Fung and Megan Ramos, sharing the joy, anguish, and humor of what really happens during a fast. Through her raw transparency and brash sense of humor online at FastingLane.com, she has inspired many to find their own health and hotness.

MEGAN RAMOS is a Canadian clinical researcher and expert on therapeutic fasting and low-carbohydrate diets, having guided more than fourteen thousand people worldwide. Passionate about overcoming her own health hurdles through fasting, diet, and lifestyle modifications, she co-founded Intensive Dietary Management and The Fasting Method. Her days are dedicated to creating education and providing support to those looking to regain control of their health. Megan lives in Toronto, Canada, with her husband and two greyhounds. She is a director of the nonprofit organization Public Health Collaboration Canada (PHC Canada) and works on the editorial board of the *Journal of Insulin Resistance*.